DIGITAL SUPPLY CHAIN MANAGEMENT

DIGITAL SUPPLY CHAIN MANAGEMENT

Driving Resilience in Supply Chains Post COVID-19

J.P. Saxena

Ph.D., M.B.A., D.L.L., D.I.M., B.E. (Elect.), B.E. (Mech.)

Formerly
Professor, Amity University, India
Consulting Engineer, Saudi Cement Company, K.S.A.
Director R&D, National Council for Cement & Building Materials, India

ISBN-13: 9780578367316

Dedication

This book is dedicated to my wife, who has left for heaven, the ultimate destination. However, she has always been with me during the writing of this book

PREFACE

This book is written continuing the threads of Industry 4.0 to focus on the digitalization in supply chain management and helping in enhancing the supply chain resilience. Digitalization has changed the way in which supply chains are managed and how the digital technologies contribute to improving resilience of the supply chains. The timing of the book has been concurrent to unfortunate COVID-19 onset, which impacted the life all over the world with extreme disruptions never seen in the history of mankind. Starting from China, COVID-19 rapidly spread to other countries. Even the advanced countries like USA, UK, and European countries were not spared though these countries are known for their good healthcare systems.

COVID-19 took every country by surprise and exposed their unpreparedness to meet the pandemic or disasters. It led to mass-scale deaths, lockdowns; cancellation of international flights, and trains; closure of working places, restaurants, places of visits, schools, colleges, imposing compulsory use of masks and following other restrictions including social distancing norms. Even the social gatherings, religious places, and marriage functions were affected by the COVID-19. There were shortages of everything including food, healthcare materials, life-saving devices, and drugs.

Supply chain management emerged as one of the key factors in the management of even the normal daily life. Shortage of manpower, transportations systems, and materials were noticed everywhere. People adopted online purchasing as partial substitute to their usual shopping habits due to restrictions on their movements due to forced circumstances. This was the time when importance of digital technologies in effective management of supply chains to meet basic needs of life was realized maximum. Forced by the difficult situation and experiencing greatest personal loss in life with the death of wife from COVID-19, the author took up the task of writing this book to help supply chains making them resilient enough to face the challenges of pandemics and disasters.

This book has been organized in four Sections having sixteen Chapters. Section I has three chapters followed by Section II with ten chapters, Section III with one chapter, and Section IV with two chapters.

Section I

Chapter 1 deals with the global supply chain, factors that affect the supply chain management. This Chapter further deals with the new challenges faced by the supply chains.

Chapter 2 discusses the impact of COVID-19 on supply chains. Explaining the COVID-19, the Chapter discusses the way COVID-19 spreads and how it affected various countries like China, Germany, European countries, and the USA. The Chapter highlights the impact of COVID-19 on supply chain activities. Various aspects relating to COVID-19 such as non-availability of COVID-19 vaccine, its technology, raw materials to manufacture, demand and supply, challenges in manufacturing, and other relevant aspects are covered in this Chapter.

Chapter 3 discusses the concept of resilient supply chain, disruptions from the COVID-19, different types of resilience, its definition, attributes, and the ways to build resilience of the supply chains.

Section II

Section 2 of the book is about digital technologies used in supply chains. The section comprises of nine chapters covering important technologies which include Artificial Intelligence (AI) and machine learning, robotics, Internet-of-Things (IoT), Cloud technology, Big Data and Analytics, Radio Frequency Identification (RFID), Simulation technology, software in digital supply chain management, and 3D Printing. The last chapter in this Section discusses digital transformation strategy.

Section III

Section 3 relates to Cyber Security comprising one Chapter. Chapter 14 deals with cyber security of supply chains covering definition of cyber security, classification, sources, examples of security breaches, measures to mitigate cyber risks, strategies to control cyber-attacks, and a framework for mitigating cyber risks.

Section IV

Section 4 with two Chapters is about managing major disruptions. Chapter 15 refers to SCM strategy in managing major disruptions from disasters and pandemics and discusses issues in enhancing the resilience of the supply chains to deal with risks. Chapter 16 is the last Chapter of the book, discussing the lessons learned from COVID-19, and the road ahead for the next normal of supply chain. The Chapter discusses strategic action plans on short-term, medium-term, and long-term basis.

Intended Audience

This book will be useful to all the stakeholders of supply chain management comprising practicing professionals in the industry, service sectors, educational institutes, and the government. The book is useful for managers, engineers, executives, planners, researchers, professors, lecturers, and the students of various courses like MBA, BBA, Post Graduate Diploma in Supply Chain Management and Logistics Management. It will help strategic decision makers and policy makers in all aspects of supply chain management like procurement, warehousing, inventory management, and transportation.

TABLE OF CONTENTS

SECTION I

IMPACT OF COVID-19 ON SUPPLY CHAIN MANAGEMENT

1

Typical Challenges in Supply Chain Management

Supply Chain Management (SCM) refers to the management of activities related to the flow of materials and services between businesses and locations. These activities include the movement and storage of materials such as raw materials, semi-finished goods, and/or finished goods, and the order fulfilment steps from point of origin to point of consumption.

The finished product(s) of a supplier (let's say Supplier #1) may be the raw material(s) for another supplier (say Supplier #2), whose finished product(s) may be the raw materials for another supplier (say Supplier #3). Thus, in this example, Supplier #2 is a customer of Supplier #1, and Supplier #3 is a customer of Supplier #2. The processes used to manage such a flow of raw materials and services from one or more suppliers at a point of origination to deliver finished goods or services to an end customer at the point of consumption are together referred to as SCM activities.

To understand how a network of SCM activities can rapidly become complex to manage, let's consider another example from our everyday lives, such as the process for manufacturing an automobile, such as a car. Figure 1.1 on the next page shows the image of various spare parts/components that are needed by a car manufacturer, such as Chevrolet, to build a car. Chevrolet may build some or all the components itself or may choose to buy some of these components from other suppliers. For example, Chevrolet may decide to buy the tires needed to build the car from one or more tire manufacturers (such as Michelin or Good Year or Bridgestone Tires, etc.) Considering each car is made up of thousands of components, the process to manufacture a car comprises of numerous chains of suppliers selling semi-finished goods to their customers, those customers selling their semi-finished goods to their customers, and so on.

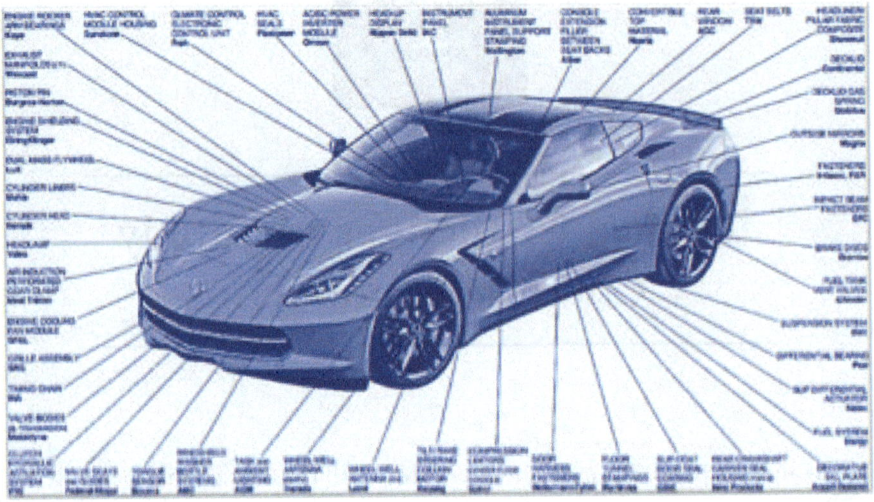

Source: Image from the internet

FIGURE 1.1: EXPLODED VIEW OF GENERAL MOTOR'S CAR

Further, most components could be built or bought by each car manufacturer from one or more alternate sources of supply that may be in different countries, time zones, and may need different lead times and cost to fulfill any orders that are placed for such parts. They may also be doing business in different local languages, currencies, may have to apply different tax amounts based on the applicable jurisdiction areas, and may have to comply with certain procedural, contractual, or regulatory requirements as well. Thus, the supply chain processes involved in the manufacture/delivery of certain goods or services to end customers could turn out to be long, complex, time consuming, and/or expensive. Management of such supply chain activities is known as SCM. Note that the examples of supply chain processes described above are not representative of all the SCM activities that are needed to manufacture complex goods and services. Management of an inventory of semi-finished goods, their picking, packing, and shipping to customer warehouses or locations using the most safe, economical, and appropriate mode of transportation are also considered supply chain activities.

1.1 The Global Supply Chain

In the case of global supply chain, materials move from one country to country through different political and commercial systems involving people of different nationality, culture, background, language, age group, and habits. Work ethics and the value system of each place differs. Most leading organizations have applied the principles of SCM over the last several decades to address these variances and challenges with the goal of meeting the end customer's requirements most efficiently and effectively. The remaining sections of this chapter describe the typical factors that have impacted SCM activities across industries and geographies over the last several decades versus the new factors that have recently emerged because of the disruption caused by COVID-19 on SCM activities globally.

1.2 Typical Factors that Impact SCM

Traditionally, various factors that influence and pose challenge to the supply chain management can be classified in the following six categories:

i. Technical factors
ii. Technological factors
iii. Economic factors
iv. Political factors
v. Cultural factors
vi. Environmental factors

1.2.1 Technical Factors

Technical specifications of a critical material or life-saving material, if not written correctly and completely, can pose a challenge in its timely procurement. Wrong or incomplete technical details delays the procurement activity as this would not attract offers from the right suppliers. Further, suppliers may be engaged in avoidable long correspondence without fruitful results. Equally important are special directions for the packaging and transportation of materials of unique requirements. Unique requirements may include maintain special temperature condition, risk from fire, special handling needs of chemicals, the leakage which may result in a disaster.

1.2.2 Technological Factors

New technologies often pose challenges in technology transfer. New developments and technologies need trained staff to handle the supply. This is especially important in case of nuclear materials, explosives, and chemicals which are to be mandatorily handled by trained persons only. Companies are to arrange either its own staff with proper training or to hire specialists from outside the organization to receive and handle the consignment of such materials. Customs clearance of such items may also be a challenge. Technological factors are especially important in case of prohibited items.

Technology status of the country makes a difference in supply chain operations. Systems and procedures in advanced countries like the USA, UK and other European countries are different from the developing and underdeveloped countries. SCM teams in the developing and underdeveloped countries do not have the latest software support and trained workforce to operate such software solutions.

1.2.3 Economic Factors

This is one of the most important aspects in the SCM. Manager, SCM ensures that the material is from the most economic source of supply while maintaining the quality and technical specifications of the materials. Knowledge of latest Incoterms is important to understand the difference between the quoted prices by the suppliers. Some common incoterms used in the procurement are FOB (Free on Board), CIF (Cost, insurance, and freight), C&F (Cost and Freight. Many other Incoterms are also used in the global supply chain, each clearly defining the responsibilities of the buyer and the seller in the supply of materials. Lack of knowledge of the Incoterms sometimes creates misunderstanding between the buyer and supplier leading to dispute. SCM teams face the challenge in analyzing the price quoted by the suppliers located in different countries. Prices quoted by different suppliers from different countries should be compared on a common platform in decision making. Purchasing analytics is an important area of study necessary to be learnt by the procurement team to make competitive procurement. The SCM team should have strong knowledge in deciding which part is to be purchased from which source of supply and still maintain the matching requirements of parts obtained from different sources of supply located at different locations or in different countries.

Use of Big-Data by the international companies enables the procurement teams to use online data to work out the optimum solution. Price difference for the same item with similar specifications from different sources of supplies located in different parts of the world is due to many reasons, for example, cheap labor cost, availability of qualified and trained manpower, low cost of power supply, difference in tax structure in the country. Although lowest cost has been the main criteria in making purchase decisions so far, but recent experience of COVID 19 has changed this criterion. Today, SCM managers make strategic decisions based on the availability of the items. The focus now has changed to the availability of the items, possibility of transportation and availability of workforce. Due to the impact of COVID 19, the certainty factor has changed to an uncertain factor. The SCM team, therefore, must find solutions within the available resources and prevailing situation.

1.2.4 Political Factors

There are different political situations in different countries. A country may be having a democratic setup of the government like in the U.S.A., India, and the U.K., where the government is formed by the elected members of the parliament of the political party in majority. The policies in the country are formulated by the party in power. If there is a change in the Government, the policies get changed. Supply chain activities are greatly influenced by the party in power. Many countries are under the control of a king. The policies of the country are decided by the king. The flexibility in the policies in a kingdom is less in comparison with the policies under the elected form of the Government. In certain countries, the government is taken over by the Military through a military coup. The policies under the control of the military are rigid. Supply chain activities are greatly influenced by the structure of the government. Each type of government has its own policies on 'buy local', non-tariff barriers, subsidies, operating conditions, and ownership conditions. Political climate of the country, therefore, greatly influences the supply chain management in the country.

1.2.5 Cultural Factors

People from different countries have different cultures and habits. Different religions also have an impact on the way people think, work, and practice business ethics. A person from one culture may not understand the feelings and the meaning of what is said by a person from a different culture. For example, the cultures of Asian countries, Middle East, and European countries differ vastly. Business communications may be interpreted differently by the people of one cultural background from that of the people of other cultures. Different accents, different skill and fluency level in the language(s) used, and even the difference in educational background could influence business communications. The SCM teams face these challenges in their day-to-day work. Even a difference in the time zones, in which the resources of two organizations operate, can introduce a challenge that impacts the net output of a business relationship between two organizations.

1.2.6 Environmental Factors

Environmental factors have become especially important in supply chain management. Suppliers must supply the products that meet the environmental norms. Walmart and Amazon, the well-known super stores in the U.S.A. with operations spread over in a number of other countries specify in their agreements with the suppliers to strictly follow the environmental norms.

1.3 The New Challenges in SCM

In addition to the six traditional areas of challenges, the new Supply Chain environment faces new challenges that include collaboration challenges, managing cybercrimes, and regulatory compliance.

1.3.1 Collaboration Challenge

Globalization of the business has created a competitive business environment. Companies are prompted to consider new options to improve their market presence, increase their profitability while extending the enhanced service to the customers. A new initiative is seen in the business environment to collaborate with other organizations to spread the boundaries of the business, reaching to new markets, and reaching to the customers with better service, ease, and cost advantages. However, collaboration poses new challenges with various questions such as:

i. Who to collaborate with?

ii. What terms to collaborate on?

iii. Which activities to collaborate on?

iv. How should responsibilities be shared between the collaborating partners?

There is a wide choice of options to collaborate with other companies, for example, whether to collaborate in the same industry, same region, and in the same country. The objective of collaboration is to build capacity, capability, increase cost effectiveness and satisfy the needs of the customers in the best possible manner. In actual practice, it is not possible to know in advance every detail about a collaborating partner/supplier. It is a challenging task to choose the right partner.

1.3.2 Managing Cyber Crimes

Large number of cyber-crimes and credit card frauds are reported these days. Protecting the interests of various parties from cyber-crimes and other frauds has become a necessity and poses a challenge. The SCM team should ensure business transactions are clear, transparent, real-time, and trustworthy.

1.3.3 Regulatory Compliance

Government laydown guidelines and issue directives to be followed by the business in its operations. Use of electricity, use of water, safety of workers in construction activities or at heights, rest houses for the workers at the factory or workplace are some examples where the government enforces the regulations to be adhered to by the businesses. Environmental protection has become extremely important these days. Supply chain management teams face challenges in meeting some of these regulatory requirements due to various reasons. The SCM teams must align itself in such a way that offers regulatory compliance.

1.4 Summary

This chapter explains the concept of supply chain management that links the supplier with the buyer. Traditionally, six factors: technical factors, technological factors, economic factors, political factors, cultural factors, and environmental factors were considered to have an impact on the supply chain management. Over time, in addition to these factors, SCM teams face new challenges in finding the right partners for collaboration, managing cyber-crimes, and regulatory compliance.

2

Impacts of COVID-19 on Supply Chain Management

Sudden outbreak of COVID-19 in many countries across the world has made a history with large number of deaths. Different counties faced the impact of the disease in varying magnitude. People were taken by surprise as nobody was prepared for this. The impact was so serious that it engulfed the whole world. Healthcare authorities were looking for solutions but found themselves totally unprepared to control the situation. Supply chain management systems were totally shattered and called for a serious immediate remodeling and restructuring.

2.1 About COVID-19

Coronavirus Disease 2019 (COVID-19) is a respiratory disease that spreads by SARS-CoV-2 virus (OSHA, 2020) (Figure 2.1). The spikes that adorn the outer surface of the virus, impart the look of a corona surrounding the virion as seen under the microscope. COVID-19 started from Wuhan, the capital of central China's Hubei province on December 31, 2019. By the end of February 2020, about 90,000 people have been affected by the virus, leading to over 3,000 deaths. The disease COVID-19 quickly spread to many other countries around the world including the USA. More than 65 countries have reported positive cases with significantly increasing numbers in South Korea, Italy, Japan, and Iran (Deloitte, 2020 a). The spread of the disease took all the countries by surprise and resulted in huge loss of lives.

Table 2.1 on the next page shows the details of the total cases reported and the number of deaths in different regions of the world (Deloitte, 2020b).

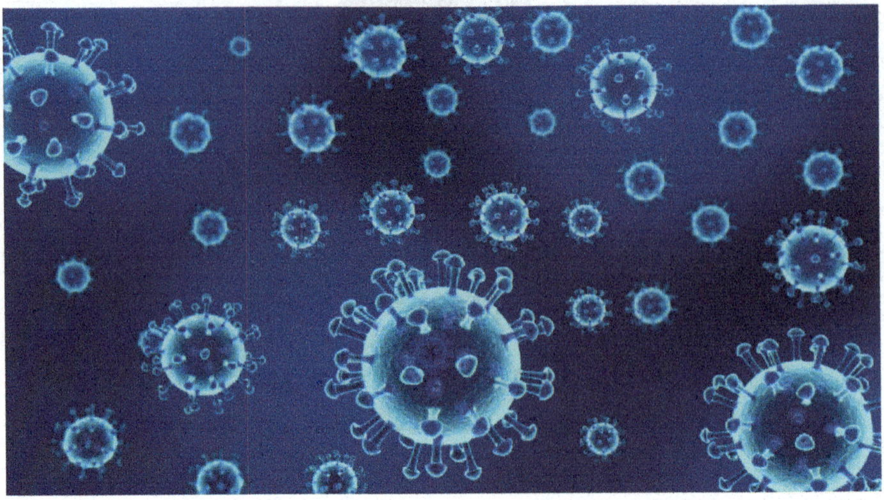

Source: CDC / Alissa Eckert & Dan Higgins

Figure 2.1: Coronavirus

TABLE 2.1: CASES AND DEATHS DUE TO COVID-19 AS OF MAY 31, 2020

Name of Country / Region	# of Cases	# of Deaths
China	>84,600	>4,600
North and Central America[1]	>2,017,100	>122,400
Asia (excluding China)[2]	>349,200	>9,700
Middle East[3]	>505,000	>12,400
Oceania[4]	>8,400	>100
Europe	>2,142,500	>180,100
Africa	>100,600	>2,600
South America	>800,800	>38,200

Source: WHO, Johns Hopkins University, McKinsey analysis

Notes:

i. John Hopkins data used for U.S., all other North America countries reporting from WHO
ii. Includes Western Pacific and South-East Asia WHO regions; excludes China; note that South Korea incremental cases are declining, however other countries are increasing
iii. Eastern-Mediterranean WHO region
iv. Includes Australia, New Zealand, Fiji, French Polynesia, New Caledonia, Papua New Guinea

Figure 2.2 on next page shows the status of the spread of the disease all over the world as on January 23, 2022. The menace of the disease has not stopped and is expected to continue for an uncertain period in the future. The latest status of the confirmed cases of COVID-19 and the number of deaths as on January 6, 2022, is shown in Table 2.2 (WHO, Jan. 7, 2022) below. COVID-

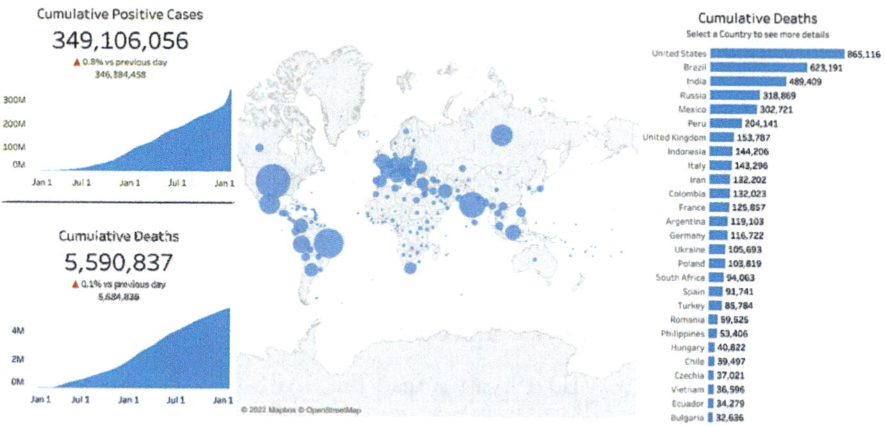

Source: Global COVID-19 Tracker, Tableau

FIGURE 2.2: STATUS OF COVID-19 AS OF JANUARY 23, 2022

19 started from China, but the most severely affected regions from the COVID-19 were Europe, North and Central America, South America, and Middle East both in terms of total case counts and deaths. The severity of the disease was extremely grave, affecting all aspects of daily life, including travel, trade, businesses, tourism, food supplies, and financial markets.

TABLE 2.2: NUMBER OF COVID-19 CASES AND DEATHS AS ON JANUARY 7, 2022

No.	Country	Cumulative Confirmed Cases	Cumulative Deaths
1	World	296,496,809	5,462,631
2	USA	56,843,327	824,175
3	Brazil	22,323,837	619,384
4	India	35,109,286	428,876
5	Russia	10,601,300	313,817
6	Mexico	4,008,648	299,711
7	Peru	2,309,856	202,867
8	Italy	6,756,035	138,276
9	Iran	6,201,467	131,778
10	Columbia	5,203,374	130,100
11	France	10,645,841	122,206
12	U.K.	13,835,338	149,284

Source: WHO, January 7, 2022

The COVID-19 crisis has changed the lives of millions of people. Imposition of international travel bans, lockdowns of the whole country, closure of shops, restaurants, fitness clinics, schools, colleges, and Universities; and cinema halls have been seen all around the globe (Deloitte, 2020b). Economies of all the countries are severely affected by COVID-19. Countries are exposed to a flood of patient loads facing inadequate hospital capacities and capabilities, and shortage of medicines. Many people lost their jobs, and many stayed at home by force. COVID-19 is, therefore, a risk for human health as well as to the social and economic situation in the affected countries. Deloitte (2020b) states that the virus is one of the greatest crises for humanity in the past 100 years.

2.2 How COVID-19 Spreads?

People infected by COVID-19 can spread the SARS-CoV-2 virus to other persons in two ways:

The infected person comes in close contact with another person (within about 6 feet). The infected person while coughing and sneezing spreads the respiratory droplets landing in the mouths or noses of the people in close distance who would have possibly inhaled them into the lungs.

The COVID-19 spreads when an infected person touches a surface or object that has SARS-CoV-2 on it and then touches his own mouth, nose, or possibly his eyes (OSHA, 2020). The symptoms of the disease include fever, cough, and/or shortness of breath. COVID-19 has also spread in people without showing the symptoms.

2.2.1 COVID-19 in China

COVID-19 started from Wuhan where 200 of the Fortune Global 500 firms have their presence. Wuhan is the transportation hub for many industries with the largest inland port in the country, and has a well-developed infrastructure in water, land, and air traffic (Deloitte, 2020a). Economists believe that the economic impact of COVID-19 on most of the industries will be limited to medium to longer-term without disturbing the overall growth. However, the consumer sector, which contributes the most to economic growth, is expected to be adversely affected with tremendous cash-flow problems due to reduced sales and high fixed costs. It is expected that the manufacturing industry will get back on track in the long-term.

2.2.2 COVID-19 in Germany

Deloitte (2020b) conducted a survey in the last week of April 2020 during the COVID-19 crisis to study consumer behavior and in-depth industry insight. More than 2,000 participants took part in the survey. In addition, interviews with experts from different industries were conducted to get the complete picture about change in consumer behavior after the crisis. Major industries affected by the crisis by forced shift in consumer behavior included retail, consumer goods, banking, tourism, and telecommunication, etc.

Impact of COVID-19 on consumer behavior in Germany has been wide ranging, which include the following:

i. COVID-19 has affected the German economy on consumption as well as on GDP.

ii. Rappeport and Smialek (2020) predict a large-scale recession of the economy with decline in economic activities within Germany.

iii. Companies find it difficult to prepare for the future due to uncertainties in easing out of lockdown measures.

iv. New sales and service solutions are being tried to secure the lost business.

v. Companies are adapting new business models such as implementation of online shops and digital customer services.

vi. Companies are reacting to changes in consumer behavior. About 13 percent of the people shifted to online buying of medicines from offline buying due to the impact of COVID-19 (Statista. 2020).

vii. Increase in contactless Girocard payments from one third in 2019 to more than half in 2020.

viii. Companies are finding it difficult to find a lasting competitive business model for the future.

ix. COVID-19 crisis affects different groups of people to varying degrees in the short and long term.

The changes in the consumer behavior are short-term due to forced new regulations or long-term triggered by restrictions. The crisis is compelling the consumers in their habits and the way of living in almost everything such as travel, shopping, sports, and meets-up. Some of the changes have long-range effects, such as purchasing from local traders, which they have never experienced earlier. Such changes in behavior of the consumers are likely to stay.

The Deloitte survey (2020b) shows that the shift in consumer behavior after COVID-19 would be in the range of 25-50 percent of the shift during the pandemic. The survey shows that consumption of the products and services varies with age and income. Lower income households do not show major changes in habits during COVID-19 crisis, but higher income households demonstrate bigger shifts in their travel and other day-to-day activities.

The data about COVID-19 infections and deaths suggests that the virus is most dangerous for older people. This influenced the behavior of young consumers, who missed social interaction and activities during forced restrictions. It is expected that older people will change their way of consumption and will return to 'old habits'.

Forced reduction in personal interactions and movements are the drivers of all shifts in consumer behavior that will have an impact during as well after the lockdowns.

2.2.3 COVID-19 in Other European Countries

In a survey conducted in August 2020 of more than 2,200 small and medium-sized enterprises (SMEs) by McKinsey (2020) in the five European countries- France, Germany, Italy, Spain, and the United Kingdom, it is found that prosperity of these enterprises has been hard hit by the COVID-19. SMEs are the lifeblood of the European economy and account for more than two-thirds of the workforce. Findings of the survey indicates the followings:

i. Revenues of the large number of small and medium-size enterprises (SMEs) in Europe have fallen since the onset of the COVID-19. Italian and Spanish SMEs have been hard hit: 30 percent in Italy, 33 percent in Spain and 23 percent in Germany.

ii. 80 percent of SMEs found the economy to be somewhat weak to extremely weak.

iii. About 1 in 5 SMEs (20%) in Europe were concerned about meeting payments obligations and retaining employees.

iv. On an average about 30 percent SMEs across Europe feared that they will have to postpone growth initiatives.

v. At the prevailing rate, about 10 percent SMEs could be expected to file for bankruptcy within six months. About 20 percent of SMEs have already applied for government support, and another 33 percent intend to apply. SMEs have the potential to become prime mover for economic growth after the crisis, but governments' support would be critical.

2.2.4 COVID-19 in the US

COVID-19 affected American life in the US very hard with loss of employment income, food scarcity, delays in getting medical care, and housing insecurity (USCB, 2020). The Congressional Budget Office (CBO), in May 2020, has given the details of the impact of the COVID-19, and the expected impact in the coming days, which include:

i. The unemployment rate in February 2020 was 3.5% which became 14.7% in April 2020. This shows the unemployment of more than 25 million people. The loss of jobs was in industries such as retail, education, health services, leisure, and hospitality that involve "in-person interactions".

ii. Smaller and newer businesses were economically affected more due to less capacity to absorb financial losses.

iii. The real consumer spending fell by 17% from February to April 2020 as social distancing reached its peak. The sale of cars and light trucks in April was 49% low in comparison to the late 2019 monthly average.

iv. Mortgage applications fell by 30% in April 2020 from April 2019.

v. Real GDP was projected to fall at a nearly 38% annual rate at growth of 5.0% in Q3 and 2.5% in Q4 2020.

vi. Real GDP was not expected to regain its Q4 2019 level until 2022 or later.

vii. The unemployment rate was projected to average 11.5% in 2020 and 9.3% in 2021 (CBO, 2020).

viii. The federal budget deficit in fiscal year 2020 due to COVID-19 was expected to be $3.3 trillion (16.0% GDP), in comparison to the January estimate of $1 trillion (4.6% GDP).

Walmsley et al. (2021) made a very systematic analysis considering three scenarios ranging from a relatively moderate event to a disaster and projected the net U.S. GDP losses from COVID-19 in the range from $3.2 trillion (14.8%) to $4.8 trillion (23.0%) in a 2-year period for the three scenarios considered. U.S. impacts were estimated to be bigger than those for China and the Rest-of-the-World (ROW) in percentage terms. The major factor affecting the results is the combination of Mandatory Closures and Partial Re-openings of businesses. These factors alone were expected to result in a 22.3% to 60.6% decrease in US GDP across the scenarios. The second major factor, considerably offsetting the overall negative impacts is pent-up demands from the inability to spend during the closures/re-openings.

2.2.5 Impact of COVID-19 on SCM

The impact of COVID-19 on supply chain activities had been very great since the day-to-day business activities as well daily life of the people was hit very-hard. Experience of all the countries around the world showed that the purchasing behavior of the people changed and shifted to online shopping to the extent possible. The demand for the products became un-predictable.

The supply chain sectors around the world on one hand suffered with no or low demand for its services and products and on other side found excessive demand for certain categories of products such as food items, toilet paper, medicines, sanitary items etc. These items were not available in the stores creating large demand on online purchasing. Lockdowns added to the problems of new government regulations that were difficult to comply with even without other complicating factors. Transportation and logistics sectors were hard hit sectors with shortage of drivers and other staff as well as shortage of transport vehicles. Virtually all the logistics sectors: road, rail, air, and water transport sectors suffered due to lockdowns. Many organizations quickly made changes to their organizational structures and business models but most of them were unable to do so. On the other hand, a sudden demand in certain sectors created booming business during COVID-19 period which is likely to continue in the post COVID-19 period also. These sectors include computers and computer accessories, and sanitary products.

The impact of COVID-19 has been so great that forced businesses to redesign their products and services, made modifications to the systems and procedures, and layouts. Safety and security issues have also taken priority over other activities. Supply chain experts need to think of new solutions in every aspect of its activities whether it is procurement, storekeeping, inventory control or transportation of goods.

2.3 Vaccine: The Biggest Challenge To Date

COVID-19 is a deadly pandemic that was responsible for 2.6 million deaths by March 3, 2021(WHO, 2020; Economist, 2021). At that time, the economic cost of the disease was estimated at US $5-$14 trillion per year (IMF, 2020; Airfinity, 2021). The top three challenges related to the supply chain of the vaccines for COVID-19 were:

i. Vaccine manufacturing processes in all the three zones (upstream, downstream and fill-in-finish) face stiff competition in the highly complex and cutting-edge technologies.

ii. The manufacturing of the vaccine is faced with the lack of availability of highly specialized equipment, personnel, and difficult technology transfer among the participating manufacturers, global supply, and manufacturing networks. The manufacturing of the vaccine needs more than 100 components, which need to be available in time for the manufacturing.

iii. Building up the additional manufacturing facilities and supplying them to centers of use is a difficult task.

Due to the severe impact of the COVID-19 all over the world in terms of number of deaths and the economic loss, developing a vaccine became the priority everywhere in the world. With the result, there have been frantic efforts in developing, manufacturing, and distribution of vaccines.

2.3.1 Status

Eleven vaccines are already in clinical use in the countries, which have approved their use. In addition to this, about 80 more vaccines are at the trial stage (Hatchett et al., 2021) and 100 vaccines are at pre-clinical stage. Simultaneously, the manufacturers and the suppliers of raw materials for the vaccines have increased their efforts to meet the suddenly developed gigantic demand in billions of doses.

2.3.2 Technology and Raw Materials

Development of technology to manufacture the vaccine and finding out the raw materials in the required quantity had been a stupendous task. The challenge from the pandemic was so big that prompted the manufacturers and the suppliers of raw materials to enhance the production capacities and search for matching requirements of raw material. It is expected that many COVID-19 vaccines may face shortage of input raw materials, packaging materials, consumables, and equipment, and therefore, may not be able to stick to the committed manufacturing and delivery schedules.

2.3.3 mRNA Vaccine: Operational Challenges

Challenges in the manufacture of the mRNA vaccine, manufactured following a chemical process, include:

i. Need for highly specialized equipment and personnel

ii. Need for consistency in manufacturing, quality, and safety of the vaccine. 100-1000 quality control checks are performed during the production process to ensure the quality of the vaccine.

iii. Capacity buildup for manufacturing and transfer of technology take long time.

iv. Need for global manufacturing network due to specialization involved. The network is due to collaborations with Contract Development and Manufacturing Organizations (CDMOs).

v. Lengthy time in manufacturing.

vi. Need for timely input supply of more than 100 components required in manufacturing the vaccine.

2.3.4 Supply & Demand for COVID-19 Vaccine

The pre-COVID-19 annual demand globally was estimated to be 3.5-5.5 billion only (WHO, 2020). Hatchett et al. (2021) reported that as of March 3, 2021, more than 270 million people in the world were already vaccinated.

According to the discussion paper for the Supply Chain Vaccine Summit 2021 (Hatchett et al. 2021), the global demand in 2021 was expected to be in the range of 10-14 billion doses based on the final aspired rate, need for re-vaccination, and boosters. There are geographical variabilities in demand. It was estimated that about 11 billion doses were required to cater a plan of two doses during the year 2021. The vaccinations were based on herd community needs.

Countries in the world have been classified in three categories: The High-Income Countries (HICs), the Upper Middle-Income Countries and the low- and middle-income countries. All the three groups of countries are keen to implement the vaccination programs, but the targeted plans are different for each group. It is pointed out in the discussion paper that the regional and global mechanism COVAX have declared secured doses of about 9-11 billion for 2021, the breakup of the secured doses for 2021 is given in Table 2.3.

The above demand projections are not going to meet the total quantity of vaccine required as the three category countries are not planning to take the full quantity needed by them. The HIC plans to take full requirements of vaccinating their population, UNICs could vaccinate approximately 35% to 52% of the population, and AMCs intend to vaccinate approximately 28% to 42% of their population.

In addition to these challenges, there were challenges related to the expiration date of some vaccines, wastage of raw materials, and unethical practices expected in distribution of the vaccines in some parts of the world. Financing the committed demand was another big challenge for the vaccination program. Uncertainties in the demand posed the greatest challenge.

2.3.5 Uncertainty in Vaccine Supply

Due to the urgency of the need, and the significant risk to the lives of human beings across the globe from further delay in making the COVID-19 vaccines available to all, these vaccines were not tested as thoroughly, and on as many humans as one would normally expect before the release of such a medicine under regular circumstances. Due to all the uncertainties surrounding the future demand of the vaccines, the supply of raw materials needed, the probability of success/failure of vaccine in preventing the spread of virus, the competitive landscape that would have to be addressed when selling these vaccines, the shipping method, duration, and related cost that should be accounted for, etc., the manufacturers of COVID-19 vaccines were hesitant to commit to manufacturing large quantities of the vaccine and shipping them globally. In addition, new variants of the virus added more confusion, and varying degrees of success achieved from the use of vaccine in different parts of the world, etc. contributed to the low confidence level associated with the results of any model used to predict supply and demand for these vaccines.

TABLE 2.3: DEMAND FOR SECURED DOSES OF COVAX IN THE WORLD

Id	Group of the Countries	COVAX Secured Demand
1	Higher Income Countries (HIC)	~5 billion
2	Upper Middle-Income Countries (UMICs)	~2-3 billion
3	Known secured doses	~9-11 billion
4	Other (n/a)	1-5 billion
5	Total secured doses	~10-14 billion
6	Low and middle-income Countries that include COVAX Advance Market Commitments (the "AMC countries")	~2-3 billion

Source: Hatchett et al. (2021)

2.4 Challenges from Supply Inputs

Manufacturing vaccines continuously requires large stocks of inputs in the form of raw materials, consumables, and equipment needed to support all three stages within a supply chain: upstream, downstream, and fill-and-finish manufacturing process. Two key challenges faced were:

2.4.1 Challenges in Forecasting Demand

Negative impact of increased safety stocks for the vaccine and other input products on the availability of the vaccine creating an artificial shortage of the vaccine and its input supplies. Increased trade and regulatory controls in essential COVID-19 supplies are likely to disrupt the supply chain.

Besides the above challenges, shortage of the single-use bioreactor bags in cell culture and fermentation is a serious concern. Other concerns of shortages include shortage of Cell Culture Media, filters, Gamma sterilization, vials, and Lipid nanoparticles (NLPs). Long-term demand uncertainties add to the problem in investment decisions. Even shortage of syringes/needles for vaccine injection, and the capacities of healthcare workers may become a big challenge to the already strained situation.

2.4.2 Challenges in the Fill-and-Finish

In addition to the manufacturing and quality control, fill-and-finish poses challenges with healthcare products such as biologics which are to be there to meet the needs of additional capacities and further additional requirements from likely variants in the production capacities.

2.5 Summary

Coronavirus (COVID-19) started in Wuhan, China on December 31, 2019. It spread very quickly, engulfing many countries in the world taking many human lives. This Chapter detailed the number of deaths due to Coronavirus. It provided details of the impact of COVID-19 in countries including China, Germany, and European countries. The impact of COVID-19 on Supply Chain activities and on the economy of many countries was also discussed. The challenges associated with the vaccine for the treatment of COVID-19 have been dealt with in great details including the operational challenges in the manufacturing of the mRNA vaccine. This was followed by a discussion on global supply and demand of COVID-19 vaccine itself and on the factors contributing to the uncertainty in the supply of vaccine.

References

Airfinity (2021). COVID-19.

CBO. (2020). Interim economic projections for 2020 and 2021. Congressional Budget Office. *Government of the USA.* 1-27.

Deloitte (2020a). COVID-19: Managing supply chain risks and disruption. *Deloitte Design Studio, Canada.* 1-20. www.deloitte.ca

Deloitte (2020b). Impact of the COVID-19 crisis on short-and medium-term consumer behavior. *Monitor Deloitte.* Issue06/2020. 1-48.

Economist (April 6, 2021). Off the Charts-The best of our data journalism.

Hatchett, R. et al., (2021). Landscape of Current COVID-19 Supply Chain and Manufacturing Capacity, Potential Challenges, Initial Responses, and Possible "Solution Space" - a Discussion Document. *Supply Chain Vaccine Summit, March 8-9, 2021.* 1-29.

IMF (2020). The role of IMF in the Fight Against COVID-19: The IMF COVID-19 Recovery Index. *Global Development Policy Center.* 1-5.

McKinsey (2020 a). COVID-19 and European small and medium-size enterprises: How they are weathering the storm. *McKinsey Global Publishing.* 1-8.

Rappeport, A. & Smialek, J. (14. April 2020) Times.

Statista (2020). Produkte, die wegen der COVID-19-Pandemie online statt offline gekauft warden 2020. Von Statista: http://de.statista.com/statistic/daen/studie/1108149/umfrage/produkte-die-wegen-der-COVID-19-pandemie-online-statt-offline-gekauf-werden/abgerufen.

Unicef (April 2021). COVID-19 Vaccine market dashboard. https://www.unicef.org/supply/COVID-19-vaccine-market-dashboard

USCB. (2020). COVID-19 pandemic in the United States. *United States Census Bureau. Government of the USA.* 1-38. http://www.census.gov/data-tools/demo/hhp/#/?measures=HIR.

Walmsley, T., Rose, A., Wei, D. (2021). The Impacts of the Coronavirus on the economy of the United States. *Economics of Disasters and the Climate Change. 5. Springer Nature Switzerland AG 2020.* 1-52.

WHO (2020) Global Vaccine Market Report 2019, 2020, 2021, https://apps.who.int/iris/handle/10665/311278

WHO (2022). *Health Emergency Dashboard*, 7 January 2022.

3

Resilient Supply Chain Management

Essential goods and services such as food, water, medicines, electricity, petrol, and diesel form part of our daily needs. A healthy, organized, and smooth operating supply chains offers these daily essential needs to the people. Timely supply of these is adversely affected in terms of quantity and quality in times of disaster. Even a smoothly functioning supply chain can encounter significant problems and disruption, causing serious threats on many fronts at the same time. SCM managers are often faced with sudden disruptions in the day-to-day operations. The reason for the disruption could be any among the earthquake, flooding of an area, accident in a factory, a rail accident, aero plane accident, explosion, enemy attack, terrorist action or spread of pandemic.

3.1 Disruptions from COVID-19

Since 2019, the whole world is struggling with the biggest crisis from COVID 19, ever seen in the history of mankind. COVID 19's impact has not yet been over; the new wave of the epidemic has engulfed many countries. In the case of India, besides the shortage of hospital beds, doctors, nurses, shortage of oxygen for patients became one of the biggest challenges. The shortage of oxygen in the hospitals further aggravated the problem to an extent that the number of deaths suddenly increased. Patients in the ICU could not get oxygen. The situation became so critical that the Supreme Court had to intervene and order to divert the oxygen supply from industry to the hospitals. The Government of India quickly moved to send oxygen cylinders and tankers through rail and airplanes. Oxygen plants were also imported from the Germany. The outburst of the problem is so sudden and large that supply chain managers were taken by surprise and faced the biggest challenge of their professional careers. A war like situation posed many questions to the supply chain managers, such as:

i. What are the primary reasons for shortages of essential products?

ii. What are the potential sources of supply of such products?

iii. Where are these sources of supply located?

iv. Can these dormant manufacturing facilities be revived to address the shortage in supply of essential products, and if so, how?

v. What are the different ways in which the disruptions in the supply chain of essential products can be assessed?

vi. How can such disruptions in the supply chain be managed?

vii. Is it possible to improve resilience in the supply chain of essential products, and if so, how?

viii. What are some of the other considerations (such as forecasting of supply and demand, inventory optimization, leveraging the supply and distribution networks effectively and efficiently, cost control, insights through data and analytics, etc.) to address such disruptions in the supply chain?

ix. How can the risks in a supply chain be identified, prioritized, monitored, evaluated periodically, and proactively mitigated?

x. What are the common challenges that emerged in the areas of resource planning and resource management within SCM because of COVID-19?

xi. What key resource management strategies should be leveraged by supply chain managers to drive higher satisfaction, retention, and productivity from supply chain resources in the wake of challenging economic and working environments, and rapid changes in social and organizational behavior due to COVID-19?

The large number of questions and their wide variety show the need to manage them well in advance. If not controlled in time, the situation can turn a disaster into a catastrophe.

3.2 The Situation Post COVID-19

In 2019, many companies across the globe were trying to cope with the rising unpredictability due to COVID-19. During a period of just one-and-a-half-year, market conditions changed considerably. Some companies faced collapse of demand within a period of just a few weeks whereas some others were flooded with sudden and continued rise in demand, difficult to satisfy with resources at command. The after-effects of COVID-19 have continued to impact the global economy. Some companies are fighting with the situation whereas others are busy with the recovery. Increased demand in many sectors has opened the gates of opportunities creating further strain on supply chains already stretched by the continued disruption. In addition, there are other factors that are adding new challenges such as building environmentally and socially sustainable business models.

Supply chains that align themselves effectively to manage disasters, disruptions, and changing needs do a great service to the society and protect the people and the property avoiding a need for big recovery operations.

3.3 Supply Chain Resilience

Supply chains are the necessities of the present-day life as they take care of the minor disruptions and changes in both supply and demand. Normally, supply chains are designed to serve the needs of the people and business in normal circumstances. The situations are very different in case of disasters. People are at great risk due to loss of life and damage to property, and industry. Everything is so grave during the period of disruption that the supply chain systems are not able to support the immediate and even long term demands to bring normalcy. COVID 19 has shown this very clearly. Even the big and advanced economies like the USA, UK, and other European countries have seen that their supply chains failed miserably to meet the challenge of COVID 19. Other disasters with relatively lesser negative impacts have also created unthinkable miseries. The supply chains must be built in such a manner that they are capable of sustaining the supply of at least essential commodities to serve the affected population.

3.3.1 Operational Resilience

Operational resilience has become a key strategic issue for the companies, in addition to the traditional focus on cost-reduction, ensuring quality, and meeting delivery schedules. The priority for the companies is to make their operations robust and flexible enough to deal with unexpected shocks.

3.3.2 Resilience Based on Holistic Approach

It is very difficult to build resilience using a fragmented, and bottom-up approach. A management team of a high-performing manufacturing unit, part of a global production network, knows the extensive efforts made by it in optimizing processes to produce the required level of production at the lowest possible cost. Expanding the capacity to add flexibility in meeting the demand will not be the correct decision if the supply chain has a bottleneck somewhere else. The additional capacity built will not be useful if the bottleneck is in the local ports. The user on the other side of the procurement deal will not be able to use the benefits of the additional capacity unless the local port is made efficient. Frontline actions can be useful only if efficiency is improved in totality across the whole supply chain covering all the sites and business functions. A holistic, cross-functional approach can improve a company's overall resilience by about 30 to 40 percent more than traditional functional optimization. (McKinsey, 2021).

3.3.3 Periodic Review of Operational Performance

Acquisition of other production facilities can increase the production capacity of a large company but will not be a perfect solution as there may be many problems in the network such as underutilized facilities, assets located far from their major markets, and performance that's well behind best in class. Each organization must review its operational performance periodically based on two key questions:

What are the key activities, resources, systems, and capabilities needed to produce the finished products, and what are the inter-dependencies between the same?

Which resources are best suited to perform these activities, i.e., should some, or all, tasks be outsourced or are there some specific tasks that should be retained by the organization for execution to protect its intellectual property or competitive advantage?

3.3.4 Definition of Supply Chain Resilience

Supply chain resilience is the ability to provide surviving capacity, and/or add new capacity, under severe strain to support the preexisting demand and supply scenario. The supply chain network or portion of the network should be capable to continue operations such as directing, redirecting, and flowing goods and services even when its important elements may not be operational.

The Department of Homeland Security (DHS) and the Federal Emergency Management Agency (FEMA), USA (DHS, 2019) define the Supply chain resilience as "*the ability of a preexisting network of demand and supply to deploy surviving capacity, and/or introduce new capacity, under severe duress. It is the ability of a network, or portion of a network, to continue moving (directing, redirecting, flowing) goods and services even when important elements of the network are no longer operating*".

Maintaining the supply of essential commodities and services like water, food, and fuel in absence of electricity from the power grid displays the resilience of the supply chains (DHS, 2019).

The service providers are empowered with knowledge of prevailing networks, including major players, critical places, interdependencies, vulnerabilities, threats, and interact with supply chain owners and operators. They can support and improve supply chain resilience and post-incident restoration. In an emergency, service providers should be aware and have understanding to communicate effectively with experts, especially with the owners and operators of the preexisting networks as may be needed.

Transportation systems connecting nodes, public facilities, people, and communication networks, can impact supply chain resilience.

3.3.5 Attributes of Supply Chain Resilience

Many factors contribute to make a supply chain resilient, which include the following:

i. Just-in-time inventory practices

ii. Information and data exchange systems for accurate and relevant information in time.

iii. Risk analysis information.

iv. Preparing a list of industry vise large suppliers.

v. Knowledge about the regulatory requirements and constraints on supply chain flexibility

3.3.6 Building Resilient Supply Chains

Building resilient supply chains is a very challenging task. The process of building resilient supply chains includes the following five steps (Figure 3.1):

i. Analysis & Design

ii. Stakeholders Interaction

iii. Implementation Plan

iv. Execute

v. Review & Control

Step 1 (Analysis & Design): Supply chain networks are complex systems with many stakeholders, nodes, and special issues. Planning the resilient supply chains should start with one commodity or service such as distribution of electricity, water, gas, vegetables, and medicines. The supply chain manager will collect information about various aspects of the supply chain from the different sources such as research papers, reports, press releases and government directives and develop intimate knowledge on a number of issues

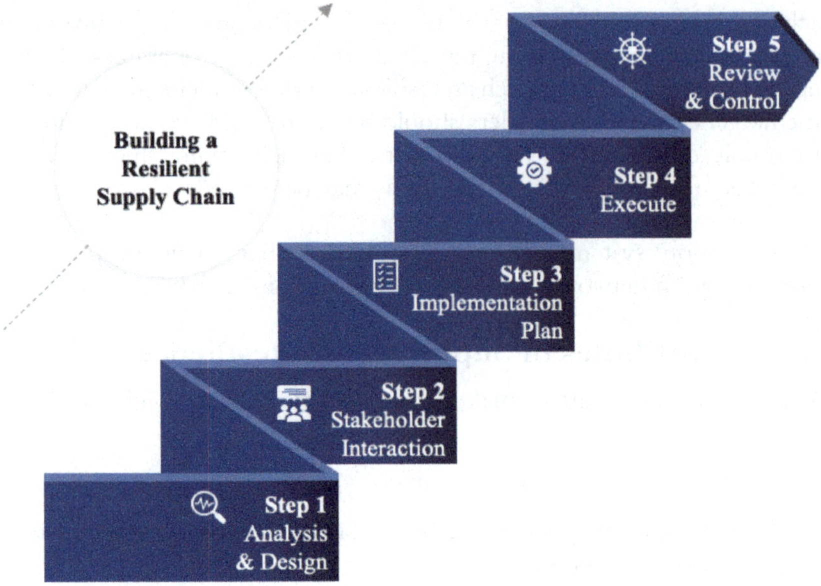

Figure 3.1: Building a Resilient Supply Chain

including interdependencies between chains. The objective is to build a strategic understanding of the supply chain and identify key stakeholders with which to interact and dig out supply chain resilience issues. Experience gained in the first supply chain can easily be expanded to other commodities and services and to community lifeline.

Step 2 (Stakeholder Interaction): Supply chain manager interacts with stakeholders and tries to develop collaborative relationships focused on enhancing supply chain resilience. Interaction with the stakeholder with different opinions on some of the issues is also carried out to narrow down the differences and develop areas of agreement.

Step 3 (Implementation Plan): Supply chain manager develops plans for implementation in consultation with the stakeholders. A common ground is created for smooth implementation of the plans. Implementation plans developed in consultation with the stakeholders have greater chances of success and help in accelerating the pace to build resilient supply chains. Supply chain managers will take adequate corrective steps in implementation of plans which are important for creating resilience in the supply chains but have not received clear agreement from stakeholders with difference of opinions.

Step 4 (Execute): This step is one of the critical steps in developing resilience in supply chains. The step includes activities for the development and implementation preparedness including tabletop exercises, with partners to enhance supply chain resilience. These preparedness activities may include actions to collaborate with private sector partners to reduce disruption of their nodes or links or to create continuity in the event of a disruption. Emergency Operations Plan or logistics plans are reviewed and revised to move according to prevailing circumstances outlining how the jurisdiction will implement logistics operations considering potential supply chain issues.

Step 5 (Review & Control): This step assesses the success or failure of the implementation plans. The plans are modified by refreshing the data obtained from the analysis as needed and continuing to collaborate with stakeholders. This is an ongoing exercise to improve the resilience of the supply chains. The step is very critical in view of changing scenarios and needs corrective actions as an ongoing process. This step identifies the need for new collaborations, development of new resources, creating new routes and transportation facilities or adding additional health care facilities. Need for new strategies and policies are identified for future improvements in resilience of supply chains.

3.4 Case Study of the State of Alaska, USA

This case study relates to 'Evaluation of Supply Chains and Resilience Building in 2012 (DHS, 2019)'.

Problem: The State of Alaska, USA is very prone to disasters with severe winter weather, volcanoes, and earthquakes that could cut off Alaska's normal supply chains for extended periods of several weeks or months. Alaska receives its supplies either by ship from Seattle, Washington or by air. A 9.2-magnitude earthquake followed by subsequent tsunami in 1964 paralyzed its power infrastructure, shipping, and distribution networks (DHS, 2019).

Analysis: In 2012, the State of Alaska analyzed the supply chains that provide food, supplies, and fuel to the state.

Findings: Based on the analysis, the State of Alaska has taken following actions:

i. Procuring shelf-stable meals for 40,000 survivors for seven days.
ii. In addition, the State identified the need to store medical supplies, cold-climate water purification systems, and generators.

iii. Storage facilities for food that was culturally and nutritionally suitable for both the urban city residents and for rural Alaskan diets, such as smoked salmon.

Result: Recent catastrophic logistics planning indicates that Alaska has seven days of food available following a disruption to the supply chain. The actions taken by the State of Alaska increased the resilience of its supply chains.

3.5 Summary

This chapter discusses the impact of disruptions from disasters like COVID-19 on the supply chains. The concept of supply chain resilience is explained with its definition. The discussion clearly brings out the attributes of resilience in the supply chain. Disaster management and resilient supply chain are dealt with, followed by a 5-step process that can enhance the supply chain resilience.

References

DHS (2019). Supply Chain Resilience Guide. *A publication of Homeland Security*, USA. 1-44.

SECTION II

DIGITAL TECHNOLOGIES IN SUPPLY CHAIN MANAGEMENT

4

Artificial Intelligence and Machine Learning

Artificial Intelligence (AI) is one of the fastest growing technologies encompassing aspects of business and has become as an important feature of an efficient business. These features include AI value, risk, trust, transparency, ethics, fairness, interpretability, accountability, safety, and compliance (Gartner, 2020) (Figure 4.1).

AI shapes the policies, practices, and principles into operational aspects of the business at the organizational and societal levels with full accountability. AI has become part of the modern business in every business activity. Its utility is fully understood and is unquestioned. Business leaders are keen to adopt, start, and develop them in their activities. All options are being tried to implement AI to gain value for the business. Substantial number of companies in the globe are changing their processes, joining hands with new partners, and searching new ways to use AI and analytics to improve performance (CIPS, 2019). It has been recognized that any change made in the organization today is going to help the organization tomorrow.

4.1 How does Artificial Intelligence work?

Machines, based on the learned experience, adjust to unique situations, and conduct various tasks such as visual perception, speech recognition, decision-making right from data collection to analysis including translation between languages in a similar way like human beings do (CIPS, 2019).

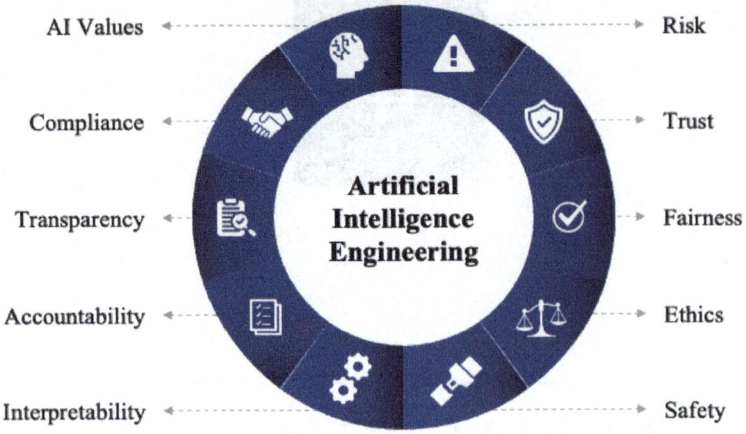

Figure 4.1: Features of AI Engineering

Procurement is one of the key activities in every organization and an important process in supply chain management. AI helps in automation of invoice processing, approving the recommended purchases and conducting various other tasks including a wide range of unpredictable tasks in procurement. Managing the supply chain disruptions, supplier compliance issues and instances of fraud are some other issues conducted by AI.

4.2 Adoption of AI by Industry

To understand the status of implementation of AI in industry, MIT SMR Connections organized an online global survey during June and July 2019 (SAS, 2020). The survey included 2,280 professionals at C levels, board-members, or management roles from a broad range of functions across 110 countries. Response of these industry-experts was obtained to find out how the drive to implement AI is affecting the organizational culture, technology strategy, and technology governance. The survey showed that:

i. Majority of respondents have made a beginning in implementation of AI technologies and are in the planning stages or have already piloted or implemented the technology.

ii. Sixty two percent of respondents informed that there was an increase in spending in the areas of machine learning in the past year.

iii. Forty percent respondents increased spending in the areas of 'Internet of Things.'

iv. Increased spending drive to implement AI is changing the organizational culture and processes creating new milestones for Chief Information Officers (CIOs) and other technology leaders.

v. Implementation of AIs affects the organizations' perceptions of AI risk and ethical issues, and how those attitudes are affecting technology governance.

vi. The extent of implementation of AI has been different in different organizations. About 5% participants are implementing AI widely across the organization, while 18% have implemented it in a few processes, and 19% are running pilot projects.

vii. AI is found more effective in comparison of other emerging technologies. Sixty-three percent of respondents said they expect AI to drive dramatic or significant organizational change.

viii. AI has an impact on software development and deployment processes. 61% of the respondents considers that AI drives significant or dramatic changes to software development processes, with 57% expecting it to similarly influence software deployment processes.

4.3 AI Case Studies in Digital Supply Chain

4.3.1 Case Study I: RABOBANK

Rabobank used different formats for customers to seek mortgages. The format was designed based on if the customers visited a branch, contacted a call-center, or logged on a website. Rabobank linked the three sets of data. Customers were helped to complete the application process. This helped sales teams to focus on contacting customers needing support and resulted in increasing the sales. Highly encouraged with the results of this approach, Rabobank adopted the approach to enhance sales and increase customer experience for 25 different products (PA and CGF, 2018)

4.3.2 Case Study II: OCADO

Ocado, one of the leading online retailers in Britain, has fully automated its operating model. Ocado designs its own AI and Robotics automation system that runs successfully (PA and CGF, 2018).

4.3.3 Case Study III: SAS

SAS's Digital twin technology uses cloud-connected sensors embedded in the machines to upload real-time operational data of the real-world machines. Manufacturers employ analytics to analyze and based on real-time operational data, evaluate the performance of their production in the field. They can monitor the operational performance of their products and can take corrective actions that may include improvement in design, innovation, efficiency, and manufacture. This capability enables them to contact the end user of the product and plan to repair or keep avoiding any disruption of highly costly breakdowns (SAS, nd). Analytics is the main driver in avoiding costly disruptive breakdowns especially in critical machines such as airplane engines, turbines etc. A heavy-duty gas turbine can be examined by looking into historical failure data and can predict the time of failure in future leading to improved maintenance plans and avoiding catastrophic failures. GE, a virtual leader in twin technology, uses a combination of artificial intelligence-driven analytics and visual sensors on matchbox-sized robots to search for cracks inside an operating machine (SAS, nd). With advances in the technology, cameras can detect cracks on even dirty or rusty turbine blades.

4.4 Summary

The chapter highlights the features of AI such as AI value, risk, trust, transparency, ethics, fairness, interpretability, accountability, safety, and compliance. The working principle of AI is explained along with the growing application in the industrial processes. An online global survey, conducted by MIT SMR during June and July 2019, has shown a wide range of impact of AI on the organizational culture, technology strategy, and technology governance. Three case studies presented in the Chapter show the success stories of the companies with the implementation of the AI.

References

CIPS (2019). Digitalization in Procurement and Supply. *Chartered Institute of Procurement & Supply*. Melbourney.1-20.

Gartner (2020). Top Strategic Technology Trends for 2021. *Gartner Inc.* 1-13.

PA and CGF (2018). AI and Robotics Automation in Consumer-Driven Supply Chain: A rapidly evolving source of competitive advantage. A publication of *PA Consulting and Consumer Goods Forum (CGF)*, London, U.K. 1-54.

SAS (nd). Modern manufacturing's triple play: Digital twins, analytics, and the internet of things. *SAS Institute Inc in the USA*. 1-3.

5

Robotics

Robotics is one of the major technological trends influencing the life of people in a positive way. Innovative robots are being inducted in various sectors like hospitals and industry. Robots work in collaboration with the doctors and nurses in the hospitals and with workers in the factories and industrial plants.

Robot is a machine developed to carry out complex actions as planned and programmed. The word robot originated from the Slavic word robata (Wikipedia, 2018; Jain, 2019). William Grey Walter developed the first electronic autonomous robot in Bristol, England in 1948 (Wikipedia, 2018). Jain (2019) argues that a machine to be termed as robot should be electronically programmable, capable of processing data or physical perceptions, able to work autonomously, able to move, able to perform certain functions and processes, has the flexibility to adjust according to the environment, and can work intelligently like human beings.

Robots have contributed extensively to the industry and have made an impact on the daily life of the people. They interact with one another and work safely with humans. Robots learn from humans to perform complex and dangerous tasks (Rouhianen, 2018). Robots do works that are dangerous, repetitive, monotonous and offer improved skills with accuracy, precision, and strength (DPDHL, 2016). Robots improve our capacity to do more work and enhance productivity. A good example of the dangerous use of robots is a drone used in military operations. Initially, robots were used for entertainment purposes but there was a significant shift in their use with the industrial revolution (Industry 4.0) later.

5.1 Uses of Robots in Different Areas

Over the years several types of robots have been developed which look and act like humans. Table 5.1 shows some of the areas of application and the purposes for which robots have been used.

TABLE 5.1: USE OF ROBOTS IN DIFFERENT AREAS

Application Areas	Purpose of Robot
Healthcare	For surgery, guided vehicles, and lifting aides.
Home	Vacuum cleaner, lawnmowers, and other domestic chores
Military Operations	Bomb disposal and transportation, drones
Law enforcement activities	Traffic control
Children's entertainment	Toy dinosaurs, robots for hobby/competition purposes
Space explorations	Example: Mars Rovers

These days robots are available in a variety of shapes, sizes, and mobility capabilities, and include stationary robots with robotic arms with limited movement, wheeled robots. legged robots, flying robots, swimming robots. Today, even human-shaped robots have been designed as companions. Pepper (SoftBank Robotics, 2019) named robot has been developed based on human emotions. Some stores in Japan used this robot which is considered a domestic companion. Pepper has the capabilities to recognize emotions based on facial expressions, body movements and words spoken. The robot reacts based on the emotion it recognizes.

Robots have yet to achieve the level of making decisions on their own without programming (Rouhianen, 2018). Casey (2016) reports that a recognized example of the modern robots is a vacuum cleaner named 'Roomba' made by the company iRobot. Roomba runs on three modes: spot mode, max mode, and dock mode. Receiving the instructions from these modes, the vacuum cleaner can clean a small area in a very efficient manner and is able to adjust its work like reversing and changing the path if any obstruction comes in its way (Rouhianen, 2018). This shows that robots have an element of intelligence needed to do some specific work or operation.

5.2 Robots in Logistics

Robots have a wide range of applications in the logistics operations. Depending on the extent of automation in the warehouse, the selection of the suitable robot can be introduced to improve efficiency and productivity in the operations. Some of these robots are discussed here.

5.2.1 Goods-to-Man Picking Solution

Automated Guided Vehicles (AGV), an ASRS goods-to-man picking solution is a robot that uses barcodes and QR codes to navigate the warehouse. This system improves efficiency and accuracy in storing and picking of goods (Khuan and Raman, 2018). Amazon acquired Kiva Systems (now known as Amazon Robotics) in 2012. It gained competitive advantage through improved order picking productivity, reduction in labor and reduction in training time. Use of robots also improves product inventory control and increases order accuracy.

DB Schenker Logistics implemented the Swisslog's CarryPick system (Figure 5.1 below), an Automated Guided Vehicle (AGV), a goods-to-man picking solution. The system was combined with an integrated parcel delivery system. The modular CarryPick goods-to-man system uses low-profile self-driving robot vehicles (AGVs guided by QR codes on the floor), which drives the underneath mobile racks and delivers them to the picking workstations (Khuan and Raman, 2018). Workers pick and place the requested items in shipping boxes from there. This implementation resulted in a reduction of real estate by 20% and workforce by 65% (Khuan and Raman, 2018).

Source: Swiss log

Figure 5.1: Goods–to-picker concept

DPDHL (2016) argues that one of the biggest challenges faced by the logistics industry is shortage of high-quality manpower to transport goods from suppliers to customers. This is due to the increased demand for the more logistics workers and from the explosion of e-commerce activities that needed more parcel shipments. The COVID-19 forced various countries to lockdowns that shifted consumer preference to e-commerce. The challenge due to expected shortage of skilled workforce could be substituted by automation that can help the workers with improved productivity.

5.2.2 Trailer and Container Unloading

Many goods are shipped in ships in the oceans from one country to another country being the most economical mode of transportation. Standard size shipping containers are employed for transportation of the goods. Most of these goods are kept on the floor of the container and stacked up to the height of the ceiling. Pallets are not used to save space and load the container to transport a maximum quantity of goods. The container is loaded onto a truck on arrival at the receiving port and then sent to a distribution center. The material on arrival at the port is typically unloaded manually, sorted, and stacked onto pallets to be stored in the warehouse. The process of emptying the shipping container manually and sorting out of materials is very labor-intensive involving operations for several hours. Many long-haul parcel trucks are similarly loaded with materials from floor to ceiling without using pallets and unloaded at the receiving end port using lots of labor.

Source: DPDHL (2016)

Figure 5.2: DHL Parcel Robot

DHL in association with its business and research partners developed a new prototype – the Parcel Robot in 2003, which included of components: a chassis, a telescopic conveyor belt, a 3D laser scanner, and a gripping system made up of an articulated robotic arm and a grabber (DPDHL, 2016). The robot placed in front of a container to unload uses its laser to scan all the boxes kept in it. Analysis of the various sizes of parcel is carried out with the help of an integrated computer to determine the optimal unloading sequence. The robot picks up a box and places it onto a conveyor that transports the item out of the container and into the sorting center. Figure 5.2 shows a DHL Parcel Robot.

Wynright, a US company, is producing the truck unloading robot like the DHL Parcel Robot but with low-cost cameras to locate the boxes in place of a costly laser scanner. The capacity of the robot is unloading over five hundred parcel boxes per hour onto an extendable.

5.2.3 Stationary Piece Picking Robots

A picker, in a manual warehouse of Amazon, may be walking about 7-15 miles in a shift (DPDHL, 2016). This shows that lots of time is spent in moving within the warehouse. Amazon took the help of Kiva, a manufacturing unit of robots, to build a robot to avoid wastage of time spent in movement within the warehouse. The robot can pick up a shelf of goods and bring the entire shelf to the picker, who is at one place. After the picker has identified the needed items, the shelf moves away, and another shelf arrives to take its place. The concept 'goods-to-picker' is used in many technologies on the market today. One example of this is Swisslog´s CarryPick mobile system. It is estimated to save up to 50% of warehouse picking labor through the elimination of walking (DPDHL, 2016). Most of these systems require high capital investment due to the need for a network connecting shelves, tracks, robotic shuttles, elevators, and conveyors.

5.2.4 Pick it Easy Concept

Knapp and Viastore developed robots with the concept 'pick it easy'. Knapp's system[2] (see Figure 5.3 below) can automatically select its vacuum cup gripper to match the product. The Viastore's system also picks up items but in addition can also put the item into a final shipping carton, an added advantage.

5.2.5 Mobile Piece Picking Robots

This robot is opposite to the goods-to-picker robot system. The robot goes through warehouse shelves and picks items just like a human being. There are several companies in the field of mobile piece picking robots. One of the organizations manufacturing this type of robots is IAM Robotics, a US-based company. The company makes an autonomous mobile manipulation robot (AMMR) to optimize the picking process. Picking process is the most time-consuming task that directly impacts customer satisfaction. The company has developed the technology 'I AM SWIFTTM. Adapting the integrated obstacle detection technology, Swift goes through multiple aisles to move

Source: Knapp

Figure 5.3: Pick it Easy

safely and accurately. The robot is shown in Figure 5.4. It picks and transports products downstream at up to human-level speeds to scale operations more cost effectively. Robot fleet can be built using Swift as a standalone robot, in combination with Bolt™ autonomous mobile robot (AMR) or integrate Swift with other automation systems.

5.2.6 Co-Packing and Customization

Retailers give incentives to customers to buy items by offering very attractive discounts or by offering things like 'buy one take one'. These offerings mean putting additional stickers, building the displays, repacking the items involving a lot of additional labor, space, and time. Retailers pass on these responsibilities to the suppliers. Last-minute requests for modifications are called co-packing or customization. Customization needs opening of the items, putting stickers, repacking the items including putting them in the boxes or cartons. Such activities can be done with the help of a collaborative robot. Baxter is one such robot which can work safely with persons around. The robot is built with two arms made of rubber joints having springs and facility to shut off if it meets any obstruction. Baxter can do the work safely in a similar way as a trained person. The robot has arms to grab and can-do simple tasks performing them repetitively. The technology is under development and would be fitting to the needs of the logistic robot.

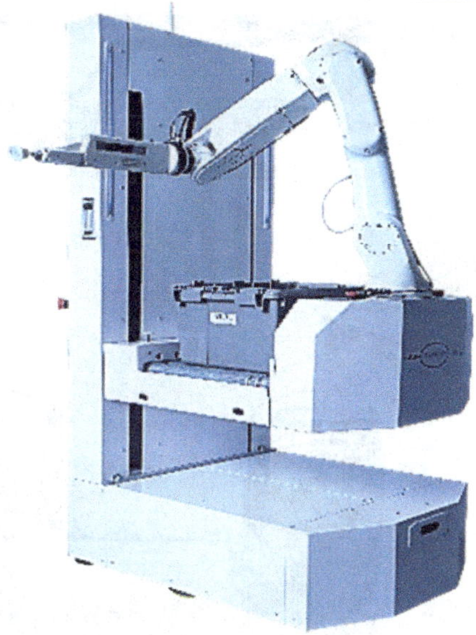

Figure 5.4: Mobile Piece Picking Robots

5.2.7 Home Delivery Robots

A new innovative delivery system has been developed to change the delivery of home services. Old people, patients, busy employed persons not able to find time to go to the store to pick up items of need, now have a facility to fall back to meet their need and get essential goods, food, medicines etc. at home without hassles. Delivery robotics are now a reality and serve the people who need their services.

Dax-The Delivery Robot: One of the delivery robots is Dax urban interface delivery robot (Figure 5.5) (Dax, 2020). DAX can use sidewalks to navigate a senior living facility. Dax is unique as it is built to interface with people. Dax is the only interactive urban robot that uses a "neck" and facial expressions to interact.

Dax operates as a semi-autonomous delivery robot, but it can do many functions specific to different industries and research projects using smart chips to make smart decisions about things such as where to drive, when to stop, and a live video feed. It has transformed the quality of life. The robot is safe on the road as it has collision detection features. Along the way Dax's cameras and sensors help it avoid obstacles and cross streets safely. Dax is equipped with chips that help it make smart decisions to reach the target destination. Dax robot. can go anywhere that is ADA compatible. Dax is designed for a max speed of 4 mph and has a turbo boost that can be used for crossing streets. Dax also has a built-in emergency stop in case it needs to

Source: Daxbot (2020)

Figure 5.5: Dax Home Delivery Robot

be reined in. Dax is 3 feet high and 2 feet wide. Three feet robotic can carry 537 Ping Pong Balls or 2 medium pizzas plus drinks. Dax has a 2.5-mile delivery range but can be optimized for longer deliveries. Dax can maneuver in both snow and rain, as well as temperatures as low as 200C and up to 400C.

5.3 Benefits of Robots

Autonomous robots offer several benefits to the enterprises and the employees (Deloitte, 2017). These benefits can be classified in two categories: primary benefits and secondary benefits (as shown in Figure 5.6).

5.3.1 Primary Benefits

Primary benefits include several benefits as discussed below:

i. **Higher Revenues:** Robots can work continuously for 24 hrs. x 7 days for the whole month or year without fatigue. They work at high speed and produce more output with less resources. Improved efficiency and productivity generate more revenues and profitability to the enterprise.

ii. **Reduced Errors, Rework, and Risks:** Humans make mistakes, robots do not. The need for rework is reduced with less wastage of men and materials.

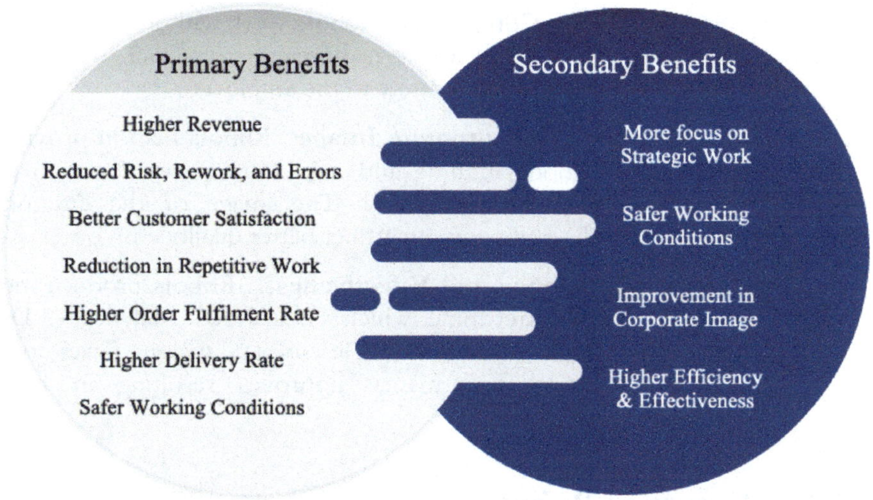

Figure 5.6: Primary & Secondary Benefits of Robots

iii. **Safer Working Condition:** Humans are at lower personal risks as the work in hazardous areas can be done safely by robots.

iv. **Reduction in Repetitive Work**: Repetitive tasks are monotonous and can be carried out by robots without losing the quality and quantity. Robots are not affected by repetitive work.

v. **Higher Order Fulfillment Rate**: Robots finish the work successfully without error or need for rework. They work at high and constant speed and produce more work within the same time in comparison to humans.

vi. **Higher Delivery Rate**: As the robots do work at higher speed without rejection, they make fast delivery.

vii. **Better Customer Satisfaction**: Customers get superior quality products as per committed delivery schedule. Customers can do better product planning in their organization.

5.3.2 Secondary Benefits

There are number of indirect benefits of robots (such as the ones shown in Figure 5.6 above) which are not visible:

i. **Greater Focus on Strategic Work**: Managers become free from routine work as the same is done by the robot. They have more time on strategic issues that help in organizational growth and the growth of the employees.

ii. **Safer Working Conditions:** Robots work without any problem in hazardous areas and relieve humans from such work. This provides better personal safety to the employees.

iii. **Improvement in Corporate Image:** Robots help in providing delivery of goods in time and good quality free from errors. Customers are more satisfied. The image of the enterprise improves. The company can attract better quality employees.

iv. **Higher Efficiency and Effectiveness:** Robots produce more goods in a shorter time which is analyzed analytically. Data produced is very useful in taking lessons and making fewer errors. The company's performance improves resulting in higher productivity.

5.4 Future of Robots

With a variety of applications in manufacturing especially in the case of automobile sector, healthcare, metals, semiconductors, electronics, food, and logistics; robots have an extremely attractive future. Chao (2016) argues that warehouse robotics would profoundly affect the logistics industry. Tractica, a market intelligence firm, shows that warehousing and logistics robot units will grow from 40,000 units in 2016 to 620,000 in 2021 (Tractica, 2017). Reese (2017) expected robotic shipments expected to reach $22.4 billion in global market value by the end of 2021, up from an estimated $1.9 billion in 2016).

5.5 Case Studies of Robots

5.5.1 Case Study I: Amazon

Delivery of a 2.3 kg package at 50mph by drone has been tested by Amazon in the UK. These drones would automatically locate the recipient using mobile phone data. It is estimated that 400 million deliveries will be made by drones in a year.

Amazon obtained a patent for an airborne fulfillment center such as an airship or blimp floating at an altitude of 45,000ft, and dispatching a drone when an order is placed (PA and CGF, 2018). Large number of products will be stocked on the airship.

5.5.2 Case Study II: Rodon Group

Rodon, an American injection molding factory, uses Baxter robots supplied by Rethink Robotics for packing the parts coming out from the production line (PA and CGF, 2018). The packing activity is usually monotonous and is carried out in a dusty environment. Workers do not like to work in a monotonous and dusty environment. Rodon offers parts at lower rates in the American market in comparison to its competitors, the Chinese companies.

In addition, Rodon conforms to the US laws and regulation in manufacturing the parts. Rodon factory is fully equipped with robots offering fully automatic production facilities operating 24 hours, 7 days a week during the year.

5.5.3 Case Study III: Boston Dynamics

Boston Dynamics is a leading robotics company with the most advanced robots. Google (Alphabet Inc.) acquired Boston Dynamics in 2013 (Glaser, 2017), then SoftBank acquired it from Alphabet in June of 2017 (Jain, 2019) and finally Hyundai Motor Group acquired it from SoftBank in June 2021. The company has designed three known robots: Handle, Spot, and Spot Mini.

i. **Handle**: The Handle model of robot is a human-like that combines wheels and limbs with lots of strength and mobility. The robot can move quickly on flat surfaces with the help of wheels. The legs enable the robot to handle nearly any terrain. Handle is of 6.5-foot height and can make 4-foot jumps (Mannes, 2017). The robot having 10 actuated joints can lift heavy items and maneuver in all kinds of spaces. The robot is equipped with electric and hydraulic actuators and can travel 15 miles on a full charge at a speed of 9 miles per hour. Handle was first featured in a video at the 2015 DARPA Robotics Challenge.

ii. **Spot**: Boston Dynamics introduced Spot, a robotic dog to deliver packages, at the TED 2017 conference. Boston Dynamics has used Spot to deliver packages to employees' homes to help refine its abilities and design (Glaser, 2017). With 12 joints, electrically powered, and hydraulic actuation, Spot has been designed to do both indoor and outdoor tasks and can run for 45 minutes on a single charge.

iii. **Spot Mini**: Spot Mini is a smaller version of Boston Dynamics' robot dog to work in office, home, and outdoor environments. Spot Mini can carry up to 30 pounds, has 17 joints, and can climb stairs (Jain, 2019). Spot Mini weighs 25 pounds and can operate for 90 minutes on full charging. The Mini has an arm that can be used to pick up and deliver objects. The robot is equipped with stereo cameras, depth cameras, and position/force sensors in the limbs that help with navigation and mobility.

5.6 Summary

Robots have been developed to do complex actions as planned and programmed. The evolution of robots has been discussed in the Chapter. Robots are used in many areas with different uses. This Chapter highlighted the use of robots in healthcare, home, military operations, law enforcement activities, children's entertainment, and space exploration. Robots are extensively used in logistics to improve the efficiency and operations which include Goods-to-man picking solution, trailer and container unloading, picking of stationery pieces, pick it easy, mobile piece picking, co-packing and customization, and home delivery. The primary and secondary benefits of robots was also discussed. It is projected that warehousing and logistics robots will grow in use.

References

Casey, H. T. (2016). You'll Be Surprised How Many Robots Now Clean Our Homes. *Tom's Guide*, 07 Nov 2016. [Online]. Available: https://www.tomsguide.com/us/roomba-vacuums-how-popular,news-23793.html. [Accessed 17 Nov 2018].

Chao, L. (2016). The Wall Street Journal. 06 Apr 2016. [Online]. Available: https://www.wsj.com/articles/robots-gain-traction-with-supply-chain-managers-1459936836. [Accessed 03 Nov 2018].

Dax, Nova Dynamics (2020). About Dax. *Dax In the News, www.daxbot.com/about.*

Deloitte (2017). Using autonomous robots to drive supply chain innovation. *A publication of Deloitte Development LLC, U.K.* 1-12.

DPDHL (2016). Robotics in Logistics: A DPDHL perspective on implications and use cases for the logistics industry. *DHL Customer Solutions and Innovation, Germany.* 1-37.

Glaser, A. (2017). Boston Dynamics has been using its robot 'dog' to deliver packages in Boston. *Recode, 25 Apr 2017.* [Online]. Available: https://www.recode.net/2017/4/25/15422130/boston- dynamics-robot-dog-deliver-packages-boston-ted.[Accessed December 21, 2018].

Jain, V. N. (2019) Robotics for Supply Chain and Manufacturing Industries and Future It Holds! *IJERT.* 8 (03). 66-79.

Mannes, J. (2017). "Boston Dynamics' Handle robot dominates parkour on wheels in new footage," *TechCrunch*, 27 Feb 2017. [Online]. https://techcrunch.com/2017/02/27/bostondynamics-handle-robot-dominates-parkour-on-wheels-in-newfootage. [Accessed 27 Feb 2017]

Rouhiainen, L. (2018). 101 Things you must know today about our Future. On Amazon.com.

SoftBank Robotics (2019). Who is Pepper? [Online]. Available: https://www.ald.softbankrobotics.com/en/robots/pepper. [Accessed 07 Jan 2019].

Wikipedia (2017) Wikipedia entry on Robot," 15 Dec 2017. [Online]. Available: https://en.wikipedia.org/wiki/Robot. [Accessed 15 Nov 2018].

6

Internet of Things

With the introduction of Industry 4.0, several digital technologies became popular in the business world. Internet of Things (IoT) is one of the technologies that created a powerful business environment with very strong communication links. In a short time, various types of things in various sizes and shapes came on the internet. Although, IoT concept was new in the business, it became one of the most talked about digital technologies. The technology has enabled relationships between humans, their artifacts, and the world around them. Today, computers can program and can store everything in cyber-physical systems. Social networks, Mobile platforms and apps, Big Data and analytics, cloud computing and Artificial Intelligence (AI) have integrated the business world bringing different things together. The impact of the technology can be seen in cheap sensors, smartphones in the pockets, smart watches on the wrists of everyone, self-steering cars, and many software solutions. According to Sogeti (2014), everything can be linked to human-to-machine chain (H2M), machine-to-machine (M2M), and machine-to-human (M2H) chains. Industry and other economic sectors can benefit from the advent of cyber-physical Internet-based systems. The M2M i.e., machine-to-machine communication is not limited to factories but is between all the devices and systems. Today, engines of airplanes, trains, and gas pipelines use sensors to receive up to date data using the IoT technology, very useful in decision making.

6.1 Definition of IoT

World Economic Forum (2015) describes IoT as a network of physical objects that include embedded technology to communicate and sense or interact with their internal states or the external environment.

Rose et al. (2015) enlarged the description of IoT as "*a global infrastructure for the information society, enabling advanced services by interconnecting (physical and virtual) things based on existing and evolving interoperable information and communication technologies*".

CIPS (2019) describes IoT as a system of interrelated computing devices, objects and people supported by ubiquitous sensors, offering the ability to transfer data over a network without human intervention. It helps the physical world to communicate with computers to decentralize analytics and decision making, thereby enabling real-time responses by recognizing business trends. Khuan and Raman (2018) argue that IoT is a revolution of information and communication technology.

6.2 IoT and Bullwhip Effect

IoT is different from the "Bullwhip effect," in which a manufacturer of a product experiences high variations in the demand and supply. The adverse effects of the 'Bullwhip effect' is controlled by the widely used industry practices, such as sharing point-of-sales data with the manufacturer, vendor managed inventory (VMI).

6.3 IoT - An Essential part of SCM

IoT helps the supply chain by curing the problem of biases relating to information exchange and human decision-making by using a different information and decision-making ecosystem. Okano (2017) describes IoT as internet based technical infrastructure, which enables interchange of items and services along the global supply chain network. Porter & Heppelmann (2014) point out that IoT offers new opportunities in terms of use, cross-functionality, and increased reliability.

Rong et al. (2015) point out that IoT capabilities can help companies improve the efficiency of their supply chain operations and help innovation. IoT can also be used in tracking the goods geographically, provide improved situational awareness, help in sensor-driven decision making, automate production processes, optimize resource use, and allow real-time sensing of unpredictable conditions (Chui et al., 2010). Phadnis (2015) and Ravi and Wu (2015), in respect to a chemical industry, found that IoT offers lower variability in ordered and shipped quantities, higher revenue with the same or lower finished goods inventory levels, lower work-in-progress and raw material inventories, fewer lost sales, automated procurement and production planning, improved process quality and safety. Okano (2017) points out the risks associated with IoT such as potential threats to the security of a company and privacy of the stakeholders.

6.4 IoT Framework for SC Performance

Khuan and Raman (2018) proposed a framework to explore the opportunities in improving the supply chain performance using IoT, which consists of three components: data gathering, data sharing, and interpretation and decision making (see Figure 6.1).

Data gathering includes collecting all types of data from all the sources at higher frequency. Data sharing is instantaneous without any distortion. Interpretation and decision making is done based on feedback as well as predictive inputs. The result of these three activities of data gathering, data sharing and interpretation and decision making provide number of advantages such as lower total cost, higher availability, faster response, effective responses in the face of unexpected changes, lower product spoilage, improved customization, and increased sales

6.5 IoT Case Studies

6.5.1 Case Study I: Rentokil

Rentokil, a world leader in pest control, launched an award-winning IoT cloud platform to improve services. The company rolled out connected traps fitted with a sensor that sends a message whenever a trap needs clearing to 15,000 customer sites within less than a year. The Rentokil platform was capable of handling 24 million messages a day (PA and CGF, 2018). Rentokil has capabilities for significant scalability for the future and could explore new ways of using data to transform its business.

6.5.2 Case Study II: Philips

Philips, with the help of the Internet connection, knows exact information about the hue lamps. How are the hue lamps being used and which color is the choice of the consumers? Philips was able to know what improvements could be introduced in the lamps and which colors were favored (Sogeti, 2014)? Philips has set up a connection with the customer platform. Some of the features of the arrangements are:

i. Can connect with the customers through the Internet 24/7

ii. Able to get user experience through improved human-machine interaction

iii. Established an M2M connection from the user to the platform

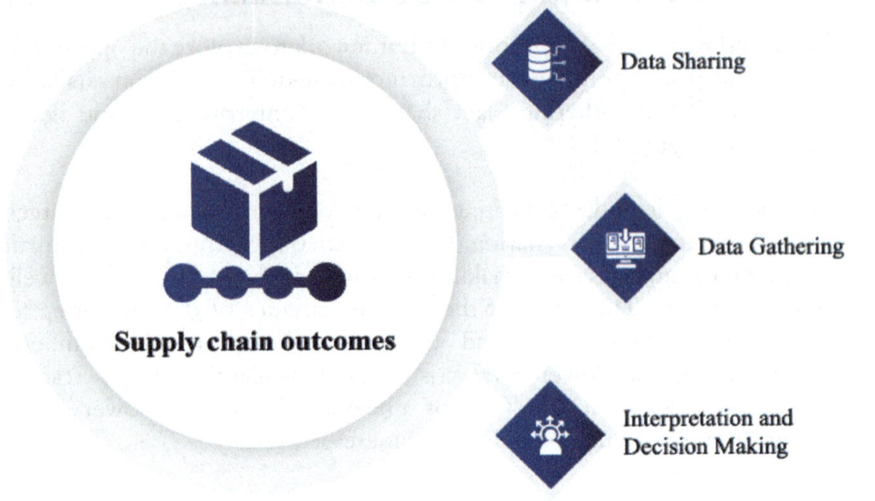

Figure 6.1: IOT framework for SCM improvement

iv. Big Data Analytics are generated on the platform adding value to the data.

6.6 Internet of Services (IoS)

Like IoT, IoS offers new possibilities for companies and private customers to easily access information via web technologies (Hofmann & Rüsch, 2017). Barros and Oberle (2012) describe service as "a commercial transaction where one party grants temporary access to the resources of another party to perform a given function with a related benefit. Human workforce and skills, technical systems, information, consumables, land, and others are the various resources to be tapped. As an example, two competing cement companies may agree to cooperate with each other wherein one of the companies may ask other competing cement company to produce its branded cement and supply the cement to the market in bags with its own logo, name, and quality standard during the period of long shutdown of its plant due to major repairs or implementing some technological changes in the plant. The requesting cement company does not want to lose its market share and keep supplies of cement under its own brand name during the period of shutdown. The requesting cement company may ensure quality of the product by its own quality control and takes responsibility for any quality issue faced by its customers. This type of arrangement between two parties, even competing ones, is termed as contract manufacturing. Hofmann & Rüsch (2017) suggest that Service-Oriented Architecture (SOA), Software as a Service (SaaS) or Business Process Outsourcing (BPO) are based on the principles of IoS.

6.7 Summary

Internet of Things (IoT) is relatively a new concept very widely accepted by industry in production and operation systems, healthcare, and others. It links the Human-to-Machine (H2M), Machine-to-Machine (M2M), and Machine-to-Human (M2H). Different authors have attempted to define the IoT technology in their own ways with the common theme that cheap sensors, smartphones in the pockets, smart watches on the wrists of everyone, self-steering cars, and many software solutions are parts of a large network of the IoT technology. The technology provides real-time information extremely useful in analytics and fast decision making.

IoT has been found very useful in several applications in supply chain management. Organizations have gained tremendously from the IoT in terms of high efficiency and profitability. The technology is very useful on one hand but on the other hand is open with potential threats and risks. The framework of IoT has three elements: data gathering, data sharing, and interpretation and decision making.

References

CIPS (2019). Digitalization in Procurement and Supply. *Chartered Institute of Procurement & Supply.* Melbourney.1-20.

Hoffmann, E. and Rusch, M. (2017). Industry 4.0 and the status as well as prospects on logistics. *Computers in Industry. 89.* 23-24.

Khuan, L.S., Raman, S.R. (Editors). (2018). *Emerging Technologies for Supply Chain Management.* WOU Press, California, USA, 1-84.

PA and CGF (2018). AI and Robotics Automation in Consumer-Driven Supply Chain: A rapidly evolving source of competitive advantage. A publication of PA Consulting and Consumer Goods Forum (CGF), London, U.K. 1-54.

Sogeti (2014). The Fourth Industrial Revolution Things to Tighten the Link between IT and OT. VINT research report 3 of 4 of Sogeti VINT.1-40.

7

Cloud Technology

Traditional supply chain has not been able to keep pace with the needs of modern competitive business environment. It lacks the flexibility needed to respond quickly to a changing business scenario. The rigidity seen in the traditional supply Chains acts as roadblock to the growth of the company. Oracle (2019) points out that many reputed organizations are faced with a typical situation in which the demands on supply chains have increased and at the same time their IT budgets have been squeezed.

The need for a modern supply chain is to have visibility into its various activities, an essential factor to manage risk, reduce cost and increase profitability.

7.1 Cloud Computing

Cloud computing is a new technology that optimizes the supply chain activities by providing infrastructure, platform, and software solutions via the internet. It involves virtualization, distributed computing, networking, and web services. In cloud computing, a program can be run simultaneously on all the connected computers in the network. Mourtzis and Vlachou (2016) explain cloud computing as a model for enabling ubiquitous, convenient, on-demand network access to a shared pool of configurable computing resources which can be obtained quickly with minimal management effort or service provider interaction.

Cloud computing helps SMEs to reduce their operational cost. Cloud computing does not need separate servers by each SME as is the case with other business technologies such as B2B, EDI etc. Cloud computing technology can move procurement and supply chain operations away from

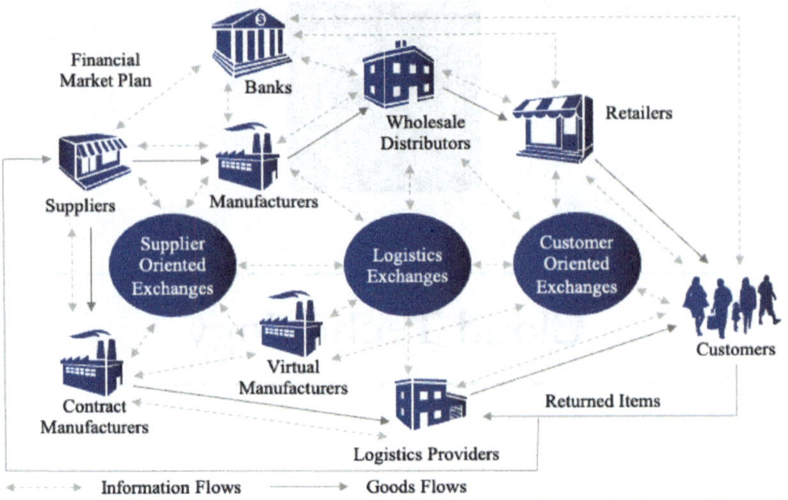

Source: Novakova and Spassov (2014)

Figure 7.1: Demand-Driven Cloud Computing Network

complex ERP built tools, replacing them with new application driven user interfaces and cloud-based analytical platforms (CIPS, 2019).

Application of extremely large, multi-party, cloud computing networks are changing the industrial scenario at a fast pace. The new technologies, such as multi-party cloud computing and the use of cycles of learning can replace the long, costly, high-risk projects focused only on internal needs by low-cost solutions to benefit the full value chain focused projects which can easily be managed with low-risk phases and each leading to fast and high returns on investments. Figure 7.1 shows a demand-driven cloud computing network that can be installed at thousands of customers deriving benefits for the entire value chain rather than serving the needs of a single business. The cloud provides the perfect architecture for multi-enterprise solutions with the integration of the internal software solutions like the ERP.

Companies can integrate the information from different stages of the value chain – from inbound logistics and production through sales and marketing thus making the system as one integrated business. This integrated information enables managers to see their value chains from end to end.

7.2 Definition of Cloud Computing

Buyya etal. (2008) define "Cloud as a system, parallel and distributed, comprising a number of interconnected and virtualized computers dynamically provisioned and presented as one or more unified computing resources based on service-level agreements established through negotiation between the service provider and consumers". Marston et al. (2011) defined Cloud computing as an IT service model in which the computing services in the form of hard wares and soft wares are delivered to customers on-demand in a self-service manner free from devices and locations.

Clients, data centers and distributed servers are part of the Cloud. Cloud based systems are simple with no compatibility problem as they use the same platform offering easy connection to every part of the supply chain. Supply chain partners can login with their ID and password to share the information in a collaborative environment with lower response time.

7.3 Benefits of Cloud Computing

Cloud computing offers number of benefits (Figure 7.2) that include the following:

i. **Better Agility**: Supply chain management in the cloud responds quickly in offering new services and capabilities with ease in scaling up and meeting the new demands.

ii. **Ease of Use**: Cloud technology enables easy and speedy deployments through outsourcing of management, and maintenance thus saving valuable time. End-users benefit from simple and pleasant experiences with accurate and timely analysis, with social interactions opportunities. The access to these can be reached at any time, from any place, and from any device.

iii. **Improved Connectivity**: Open and seamless integration of all IT solutions in various forms such as financial systems, mobile devices, RFID and IoT devices and sensors thus offering real-time information and greater visibility. This facilitates increased efficiency and productivity.

iv. **Lower Cost and Higher Efficiency**: Cloud-based supply chain management solutions offer numerous financial advantages to both the established as well as new start-up organizations. Organizations prefer low upfront investments and subscription-based pricing in place of large initial investments needed in other solutions. It is economical to arrange more affordable upgrades and new capability solutions to introduce changes with comparatively lesser expensive IT overhaul.

v. **Security**: Cloud solution offers best-in-class security features such as encryption, virus scanning and whitelist support, and enables the data center to provide physical and logical security.

Figure 7.2: Benefits of Cloud Technology in SCM

vi. **Fast, Flexible, and Safe Application**: The supply chain management in the Cloud can be changed in steps according to need. There's no need to change the entire supply chain in one go. A flexible path to the Cloud helps in prioritizing the change and implementing at the right pace for the organization in a seamless, and incremental change for an efficient operation.

Cloud computing is reliable and low-cost technology helping global sourcing and supply chain decisions at high speed. It also provides robustness to volatility with enhanced defenses for supply chain disruptions.

7.4 Service Models of Cloud computing

There are three service models of Cloud computing: Infrastructure-as-a-Service (IaaS), Platform-as-a-service (PaaS), and Software-as-a-service (SaaS) (Figure 7.3) (Toka etal., 2013). Cloud users use each of the service models for different purposes.

7.4.1 Infrastructure-as-a-Service (IaaS)

Businesses can use equipment in the form of hardware, storage space, and others as a service based on pay-per-use. According to Mell and Grance

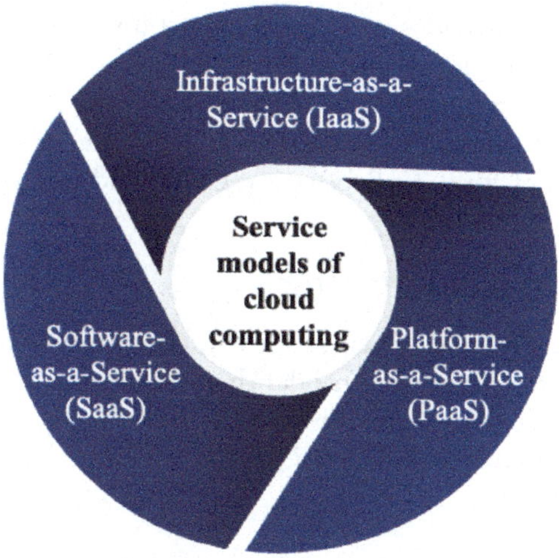

Figure 7.3: Service Models of Cloud Computing

(2011), cloud providers, under this model, offers physical or virtual machine to raw storage, firewalls, load balancers and networks. In this case, users receive the service as an outsourced service and have not to buy them as server, software, and network equipment (Conway, 2011). Amazon Cloud Services, a web-based platform, works online services like Amazon EC2 and Amazon S3 from its webpage, amazon.com (Graham et al., 2013).

7.4.2 Platform-as-a-Service (PaaS)

In PaaS, users develop and use applications, in a computing environment typically including operating system, database and programming language execution environment, offered by cloud providers (Sujay, 2011). Users do not invest in procuring and managing the hardware and software. They take virtualized servers on rent for running existing applications or developing new ones (Conway, 2011). Google Apps and Windows Azure are examples of PaaS. Windows Azure is a service offered by Microsoft. A user has the facility to build, use and manage all the applications through a network of data centers based on a Microsoft environment.

7.4.3 Software-as-a-Service (SaaS)

Cloud providers build and operate application software in the cloud. Users get into the software on various client devices with the help of a thin client interface, such as a web browser or a program interface. The cloud users are not required to manage the cloud infrastructure and platform on which the application runs but have control over the applications and possibly configuration settings for the application hosting environment (Mell &

Grance, 2011). This provides an attractive and low-cost option in place of purchasing the demanding software capabilities thus bypassing the need of applying and maintaining traditional software and hardware. Salesforce CRM, an example of SaaS, is divided into several categories such as Sales Cloud, Service Cloud, Data Cloud, Collaboration Cloud and Custom Cloud.

7.5 Cloud Manufacturing

According to Mourtzis and Vlachou (2016), cloud manufacturing (CM) is a smart networked manufacturing model that uses cloud computing with the aim of meeting growing demands for higher product individualization, broader global cooperation, knowledge-intensive innovation, and increased market- response agility.

Cloud computing has brought virtualized technologies into large-scale use. Bruque Cámara et al. (2015) say that it is not only a technical process but has made it possible to virtualize and internationalize business applications, processes, locations, and services. Bougue (2017), adding to the discussion on cloud computing and cloud manufacturing, explains cloud robotics (CR) as a technology based on the concept of cloud computing to exploit the inexpensive computing power and massive data storage capacity of cloud computing systems, combined with the ubiquitous net connectivity available, currently centered on the benefits of converged infrastructure and shared services for robotics.

7.6 Applications of Cloud Computing in SCM

Cloud computing has number of applications in effective management of supply chains (Tiwari and Jain, 2013), such as:

i. Demand forecasting and planning

ii. Sourcing and procurement

iii. Inventory Management

iv. Collaborative design and product development

v. Logistic Management

7.6.1 Demand Forecasting and Planning

Companies improve their demand forecasting and planning by closely collaborating with their chain partners like retailers, suppliers, and distributors. These cloud-based platforms obtain the data from the internet and perform analytics to make demand forecasting for all the chain partners.

The chain partners can know if the demand is increasing or decreasing and if they can handle the demand.

7.6.2 Sourcing and Procurement

Cloud-based platforms contain data from different suppliers. Companies select the supplier who meets the technical specification, offers competitive rates, and can deliver the material in right quantity and at the right time. In addition, companies and suppliers can develop mutually agreed contracts and improve their contract management.

7.6.3 Collaborative Product Design & Development

Product design and development techniques are developed with the active interaction between and across multiple expert departments of the same organization or between different organizations. Cloud technology offers a collaborative and easy platform in the development process that is shared over a secure network between different organizations. Sharing specific information, marketing firm, test result and design changes as well as customer feedback are part of cloud computing.

7.6.4 Logistics Management

Logistics management is a key area of supply chain management. Procurement, inventory control and warehousing management, and transportation including packaging are important elements of logistic management. An up-to-date and modern information system offered by cloud technology is the key success factor in efficient logistic management. Cloud computing offers many advantages, such as:

i. On demand self-servicing is available through computing capabilities without any human interaction with the service provider. Cloud computing enables use of computing capabilities through client's platforms like mobile phones, notebooks, and PCs.

ii. The cloud computing service providers pool their resources to meet the demand of various customers. Assigning the physical or virtual resources to customers is dynamically carried out without the knowledge about the location of the resources that are assigned to the consumer.

iii. Cloud computing offers great flexibility in efficiently releasing the resources to consumers according to their changing demands.

iv. Cloud computing absorbs increase or reduction in consumer demands dynamically thus enhancing the scalability of the customer.

7.7 Summary

Cloud computing has emerged as one of the most sought for technology in the digital supply chain. Cloud computing involves virtualization, distributed computing, networking, and web services. In cloud computing, a program can be run simultaneously on all the connected computers in the network. Cloud computing is a perfect solution for multi- enterprises. The chapter further explains cloud computing and discusses its definition. The benefits and applications of cloud computing have been discussed in depth. Demand forecasting and planning, sourcing and procurement, inventory management, collaborative design and product development are some of the main areas of application. It has a very good application in logistic management.

References

Bogue, R. (2017). Cloud robotics: a review of technologies, developments, and applications. *Industrial Robot: An International Journal.* 44(1). 1-5.

Bruque Cámara, S., Moyano Fuentes, J. and Maqueira Marín, J.M. (2015). Cloud computing, Web 2.0, and operational performance. *The International Journal of Logistics Management.* 26 (3). 426-458.

Buyya et al. (2008). Market-Oriented Cloud Computing: Vision, Hype, and Reality for Delivering IT Services as Computing Utilities. 10th IEEE International Conference on High Performance Computing and Communications, HPCC 2008, *IEEE CS Press,* Los Alamitos, CA, USA.

CIPS (2019). Digitalization in Procurement and Supply. *Chartered Institute of Procurement & Supply.* Melbourney.1-20.

Conway, G. (2011). Introduction to cloud computing. *Innovation Value Institute. Retrieved September* 14, 2012, from http://ivi.nuim.ie/

Graham, D., I. Manikas and D. Folinas (2013). E-logistics and e-supply chain management: Applications for evolving business, IGI Global.

Marston, S., Li, Z., Bandyopadhyay, S., Zhang, J., & Ghalsasi, A. (2011). Cloud computing - The business perspective. *Decision Support Systems Journal.* 51(1), 176–189. doi:10.1016/j. dss.2010.12.006.

Mell, P. & Grance, T. (2011). The NIST definition of cloud computing. National Institution of Standards and Technology (NIST). Retrieved September 14, 2012, from http://csrc.nist.gov/

Mourtzis, D. and Vlachou, E. (2016), "Cloud-based cyber-physical systems and quality of services", *The TQM Journal.* 28 (5). 704-733.

Novakova, Galia and Spassov, Kamen (2014) Innovative Supply Chain Technologies for Management Decisions. Conference: APDSI-ICOSCM-ASOMS 2014At: Yokohama, Japan Project: *Supply Chain Management.*1-16.

Oracle (2019). Why are supply chain leaders moving to the cloud? Publication of *Oracle.* 1 -12.

Sujay, R. (2011). Hybrid cloud: A new era. *International Journal of Computer Science and Technology.* 2(2), 323–326.

Tiwari, A. and Jain, M. (2013). Supply Chain Management in Cloud Computing. *IJITEE*, 3 (5). 2278 – 3075.

Toka et al. (2013). Cloud computing in Supply Chain Management: An Overview. Ch 12 in E-Logistics and C-Supply chain management applications for evolving business by Graham, Dcryn etal., 2013 published in *Business Science Reference*, USA. 1 - 17.

8

Big Data

Efficient operations of an organization, whether small, medium, or big, depend upon the decision-making process. Data, growing exponentially in today's technology driven environment, is a valuable resource in decision making (Giri and Lone, 2014). IDC (2011) reported that 130 Exabyte of data was created and stored in 2005 with the expectation to increase to 1227 Exabyte in 2010 and was forecasted to grow at a rate of 45.2% (Giri and Lone, 2014; IDC, 2011).

Even a small gadget like a mobile phone is a potential source of data. Every bit of data, whether it is enterprise data, personal data, historical or transactional data, is important for the organization. With the spread of COVID-19 (Corona Virus Decease 2019) all over the world, huge amount of data is generated such as the number of patients found positive, number of people found negative, number of hospitals in the region, number of doctors and nurses available to treat the patients; population of the city, state, and country; number of deaths, and number of patients treated. The exceptionally large data collected every minute, hour, day, week, month, or year is the Big Data.

The processing of large amounts of complex data becomes extremely difficult due to generation of data from multiple, heterogeneous, independent sources with complex and evolving relationships. The large amount of data continues to grow and is measured in Exabyte – one quintillion (1 followed by eighteen zeros) bytes.

8.1 Big Data

The term Big Data is used to refer to this extremely large dataset. Big Data is usually unstructured and needs more real-time analysis (Giri and Lone, 2014).

Organizations face a big challenge in dealing with this large and fast-growing mixture of data emanating from many sources. The data comes in different forms such as numbers, pictures, videos, maps, words, and phrases and makes its analysis difficult. In the business world, the data is rapidly generated in large volume through large customer transactions and social networking sites. The data generated by massive web logs, healthcare data sources, point of sale, satellite imagery is like a useful knowledge bank and needs proper storage and handling. Big companies are investing liberally in the research and exploitation of this data. Big Data technology has great opportunities in the coming days.

Data analytics and interpretation of Big Data offers a new scope of opportunities in the way of companies. Big Data offers both significant information and a way with which it can be analyzed. Big Data creates an opportunity at every stage of knowledge discovery.

8.1.1 Opportunities for Big Data

Huge opportunities for Big Data exist in the field of data analysis and interpretation in supply chain management, procurement, banking and security, communication media and services, education

8.1.2 Classification of Big Data

Big Data is classified in three categories: Unstructured Data, Semi-structured Data, and Structured data (Figure 8.1).

i. **Unstructured Data**: Unstructured data consists of information in an unorganized and complex way, such as customer reviews of the products from commercial websites, photos and other multimedia, and comments on social networking sites. Classification and separation of such data in categories is difficult but necessary for decision making. Unstructured Big Data are things that humans are saying using natural language. The formats of this type of data cannot easily fit into relational tables for analysis or querying. For example, images, audio, and video files are parts of such data. The data is loosely placed and difficult to retrieve.

ii. **Semi-structured Data**: Semi-structured data does not follow an explicit and fixed format. The data is inherently self-describing and has tags or other markers to enforce hierarchies of records and fields within the data (Giri and Lone, 2014). Examples of semi-structured data include weblogs and social media feeds etc.

iii. **Structured Data:** Structured data is organized data in categories and can easily be understood for taking decisions. Structured data is generated from network sensors embedded in electronic devices such as smartphones, global positioning system (GPS), and other available sources. Huge data generated as structured data offers new areas of opportunities such as transaction data, sales figures, account balances etc. This type of data is grouped in rows and columns within a standard database i.e., into a relational scheme. The data configuration and consistency offer it to respond to simple queries to arrive at usable information based on an organization's parameters and operational needs.

8.1.3 Growth of Big Data

The data generation is unbelievable and increasing at a very fast rate. Figure 8.2 presents the projected growth of data from the year 2010 to 2030 to reach the level of seventy-eight yottabyte (Guru99). The figure shows the share of structured and unstructured data during the period and shows there would be very large amount of data in the form of unstructured data.

8.1.4 Big Data Characteristics

Initially three characteristics of Big Data included 3Vs: volume, velocity, and variety (Laney, 2001). Shukla et al. (2015) argue that these characteristics of Big Data were suitable for the next few years only. Big Data, at an early stage,

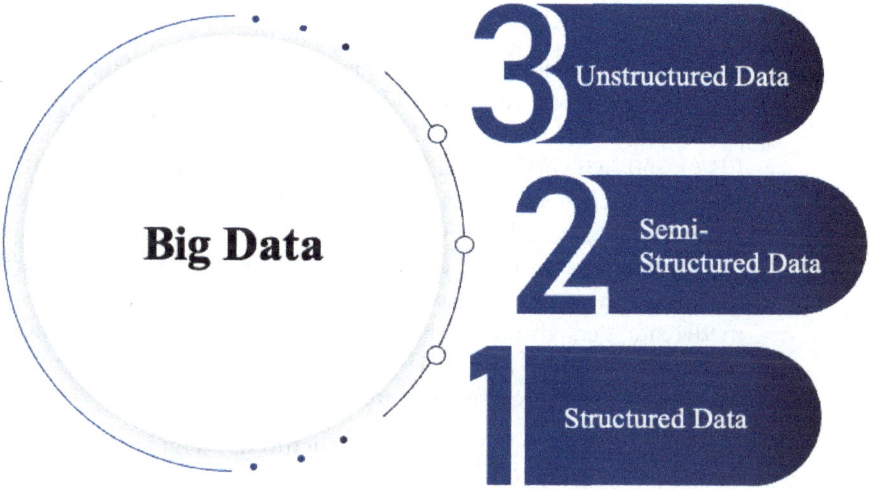

Figure 8.1: Types of Big Data

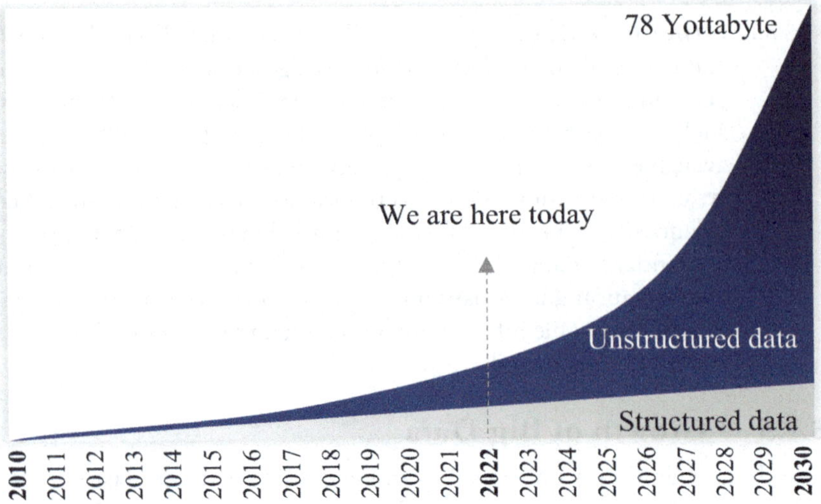

Figure 8.2: Growth Projections of Data

was basically inferred as a lot of data presented in a scattered form and needs to be processed for proper interpretation (Douglas, 2012). Later, the definition of Big Data was expanded to include 6 Vs, which included volume, variety, velocity, veracity, variability, and visualization that gave it a newer perspective (Figure 8.3) (Rijmenam, 2014). With the enlargement of the scope of characteristics of Big Data, the new enlarged scope of Big Data is not intended only for describing the amount of data, but also for the interpretation and usability of the data. Following is a brief description of each of the six key characteristics of Big Data:

i. **Volume**: Volume conveys the quantity of Big Data. The size of the data ranges from terabytes (TB) (1,000 GB or 240 bytes) and petabytes (PB) (1,000 TB or 250 bytes), to even Exabytes (EB) (1,000 PB or 260 bytes) (Giri and Lone, 2014). About one terabyte of new trade data is generated per day from the New York Stock Exchange (Guru99). A single Jet engine can produce 10+terabytes of data in just 30 minutes of flight time (Guru99). Many Petabytes of data is generated by many thousand flights per day. Facebook, the social media site, gets about 500+terabytes of new data into its databases every day in the form of photo and video uploads, message exchanges, putting comments etc. (Guru99).

ii. **Variety**: Variety includes unstructured, semi-structured, and structured data such as text, audio, video, sensor data, posts, log files and many more.

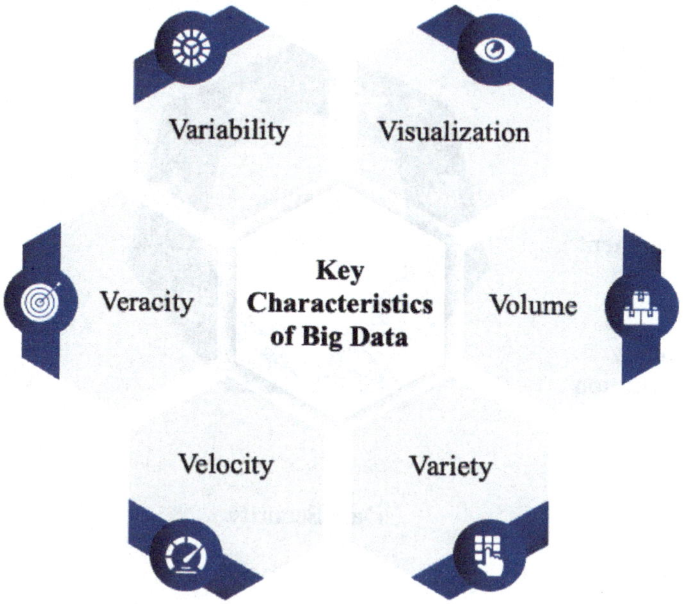

Figure 8.3: Key Characteristics of Big Data

iii. **Velocity**: Velocity means speed at which the data is obtained, processed, and analyzed. Velocity stresses on the real time processing capabilities of Big Data to meet the needs of the enterprise. COVID-19 data is processed on a real time basis to make policy decisions and take strategic decisions to control the seriousness of the diseases.

iv. **Veracity**: Veracity refers to the need for correct form of data as it is relied upon for all further analysis. It is the most important characteristic in the context of COVID-19 which has taken so many lives all over the world.

v. **Variability**: Variability of data is the same form of data but having different semantics.

vi. **Visualization**: Data visualization refers to ease of processing and interpreting to derive intelligence out of it.

8.1.5 Sources of Big Data

The main sources of Big Data, which consists of unstructured, semi-structured, and structured data, are spanned over number of sources such as social media (Facebook, Twitter, LinkedIn), sensor networks, meteorology, telecommunications, complex scientific simulations, health care systems, biological and environmental sensors, manufacturing operations, business transactions, financial transactions (Banking), transport systems (Air, rail,

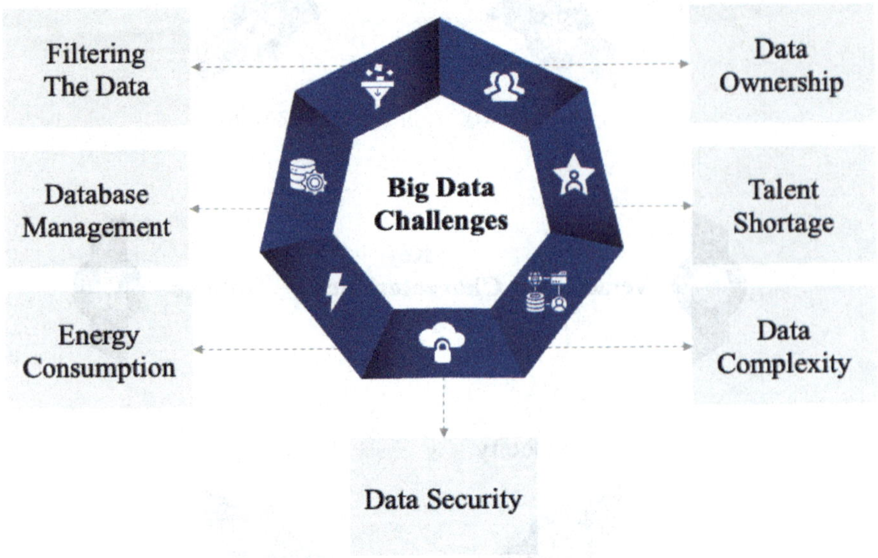

Figure 8.4: Key Challenges of Big Data

water, road systems), e-business, education and training, defense, power systems etc. (Giri and Lone, 2014).

8.1.6 Big Data: The Challenges

Organizations face number of challenges in dealing with Big Data, which is a combination of unstructured, semi-structure, and structured data, coming from a variety of sources in different forms like text files, pictures, videos, maps, images etc. Big Data poses several challenges (Giri and Lone, 2014), as shown in Figure 8.4 (see above).

i. **Filtering the Data:** It is difficult to filter out the useful data from the Big Data generated from the various sources. Filtration of the useful data poses a challenge as the process is extremely time consuming. Persons dealing with the data face a huge challenge in designing data filters.

ii. **Data Ownership:** The Big Data collected is not the private property of the person who collects the data. The person collecting the data is not authorized to use data as he likes. The correct owner of the data should be named which is a challenge.

iii. **Talent Shortage:** There is a shortage of people with analytical skills to manage the Big Data and take the benefit from the Big Data for the organizational decision-making process. Data analysis and presenting the data correctly for decision making is an ability and can be performed by professionally qualified and trained staff only. There is a dearth of such professionals posing a challenge to the organizations.

iv. **Dataset Complexity**: Big Data emanating from various sources in different forms like pictures, text files, videos, and images poses a challenge in its analysis, storage, and interpretation. As the data is often complex and in raw shape, it should be properly analyzed and presented in the proper way for future use in making strategic decisions. Different forms of data in case of the COVID-19 are a clear example of complexity of the data.

v. **Data Security**: Data has sensitive information not to be shared with those who are not supposed to get access to it. Stealing the data, hijacking the data and cyber-attacks are common and add threat to the reliability of the data. Data must be stored safely without compromising the privacy of the people. All the banks warn their customers not to share sensitive data with any person, not even with the bank officials. Organizations hire third parties to manage the data with appropriate tools to keep it safe. Data analyst is responsible to devise systems for data security.

vi. **Energy Consumption**: The energy consumptions continue to increase with the increasing volume of enormous data collected, its processing, analysis, and storage goes on increasing. This poses a challenge to the organizations due to increasing cost of energy. Organizations need to set up systems which are safe and economical.

vii. **Database Management**: The huge volume, variety and speed of Big Data is difficult to be managed with the traditional database which are designed to manage the structured data and are inadequate to manage the semi-structured and unstructured data. This is one of the most important challenges to the management. The prohibitive cost of such systems further adds challenge to the organizations. Distributed life systems (Howard et al., 1988) and NoSQL (Cattell, 2011) may be the right solutions for storage and management of Big Data.

8.1.7 Advantages of Big Data

Effective processing of Big Data offers several advantages to an organization (Figure 8.5) (Guru99).

i. **Utilization of Outside Intelligence:** Organizations can gain lots of advantages from the posts and views of the members of social media. Many marketing strategies are changed by a number of

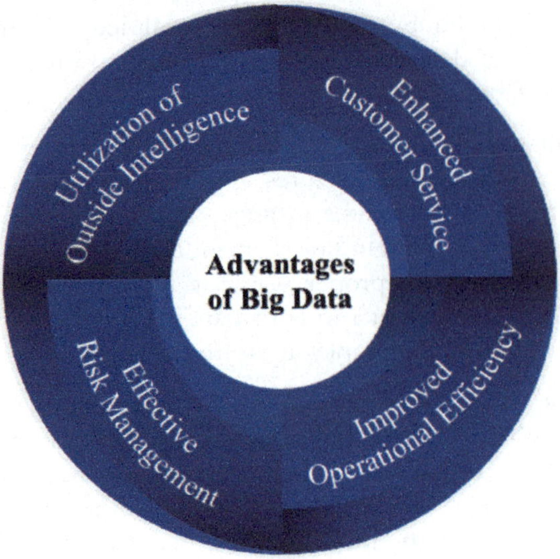

Figure 8.5: Advantages of Big Data

companies after getting the feedback from the members. This offers important intelligence from outside the organization.

ii. **Enhanced Customer Service**: Companies are serving their customers in a better way after obtaining their views about the product and services. Critical reviews and criticism help the companies to improve their products by dropping certain features. This helped the companies to be more competitive.

iii. **Effective Risk Management**: Risks can be found in advance by proper analytical investigations. Once found, risk can be controlled and managed before real damage occurs.

iv. **Improved Operational Efficiency**: Systematic processing and analysis of Big Data provide opportunities to the company to improve its operational strategies. This helps in better efficiency and improved productivity of the organization.

8.2 Big Data and Supply Chain

Big Data and supply chains are strongly connected to each other. There is tremendous growth in digital data over the years that has changed chains to meet the customer demands in a more correct manner satisfying the customer's need adequately. Supply Chain Operating Model (SCOR) can be easily related to the processes of supply planning, sourcing, make and delivery

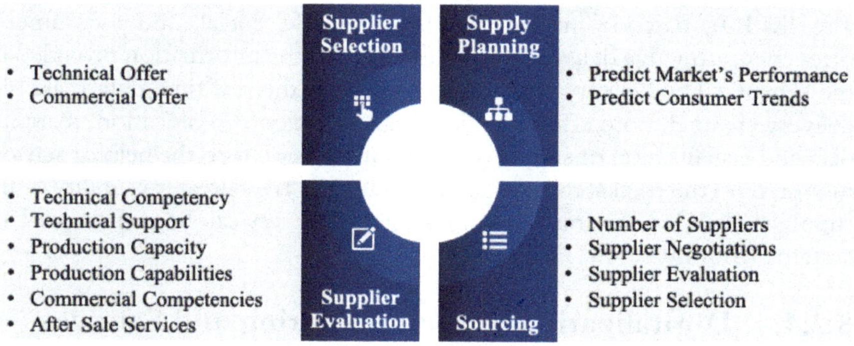

Figure 8.6: Planning Stage Processes in Supply Chain

processes. The different elements under planning stage, sourcing process, supplier evaluation, and supplier choice are presented in Figure 8.6.

Big Data through data mining and data analytics alters supply chain with the help of new data forms to solve the business problems and at the same time offers new opportunities. A good example of this is Amazon.com which serves the needs of its customers by conducting huge backend activities daily while managing queries from third-party sellers. Big Data supports supply chains in many ways such as responsive sensing analytics, forecasting the supply chain scenario, building effective and visible delivery channels, and digitalization of manufacturing and services (Figure 8.7) (Soon, et al., 2016).

8.2.1 Responsive Sensing Analytics

New advances in data and text mining, and rule-based analogies help understand the customer needs better. Digitalization has further enabled supply chain operators to understand customers' likes and dislikes better across the functions and assess the market response with the use of advance analytics. Internally setup data warehouses, database service providers, and social media are good sources of data inputs.

8.2.2 Forecasting the Supply Chain Scenario

The current method of forecasting the supply chain scenario is inflexible based on limited options to use. Forecasting the supply chain scenario choosing multiple options helps supply chain managers to consider variety of flexible solutions and decide on the best choice.

8.2.3. Building Delivery Channels

Digitalization of data and variety of digital devices provide new business solutions. Integrating data from point-of-sale, Radio Frequency Identification

Tag (RFID), barcode information from mobile, social, and e-commerce offers new attractive delivery solutions based on the information provided by the sensors. The delivery channels designed on the real time data make the delivery channel more effective, visible, and efficient. Geolocation, mapping data and visualization of supply sensing transmission with the help of sensors mounted on the trucks, rail cars, and containers offers greater visibility of the supply chain. Good examples of tracking the delivery can be seen in tracking systems from DHL and FedEx.

8.2.4. Digitalization of Manufacturing and Services

Inputs from the sensors provide real time information which is used for improved machine output and machine learning through extensive transaction of data. This input from Big Data improves the production scheduling, planning and maintenance programs with greater accuracy and efficiency. Similarly, the service industry has been improving its performance with the use of digital inputs. Hospital services extensively use sensors in monitoring the patient parameters and improve the quality of services.

The large data resources through Big Data provide new options and create new business opportunities for the stakeholders in the supply chain. The improvement in operational efficiency and profitability of companies are the result of higher speed and visibility in the various activities of the supply chain. The overall effect is shorter time taken by the product to reach the market and making the company more competitive with higher revenue returns.

8.3 Summary

Data generation is a continuous process and is increasing exponentially over the period. COVID-19 has generated huge data, difficult to manage due to multiple, heterogeneous, and independent sources. Big Data refers to this extremely large dataset. Management of Big Data has become a challenge to every organization whether in the private or public sector. This Chapter deals with various aspects of Big Data such as its introduction, opportunities, classification in three categories: unstructured, semi-structured, and structured data; growth of data, characteristics of Big Data, and advantages. The Chapter discussed the challenges faced in Big Data technology. Relationship between Big Data and supply chain management and advances in data mining have been dealt with in adequate depth. Attention is drawn to the application of Big Data technology in digitalization of supply chain management, manufacturing, and services.

References

Cattell, R. (2011). Scalable SQL and NoSQL data stores. *ACM SIGMOD Record.* 39 (4). 12–27.

Douglas, Laney. (2012). Emerging Role of the Data Scientist and the Art of Data Science. [Online] Available from: http://www.gartner.com/id=1955615 [Accessed 9th July 2012].

Giri, K. J. and Lone, T.A. (2014). Big Data – Overview and Challenges. *IJARCSSE.* 4 (6). 525-529.

Guru 99. (nd), What is Big Data? Introduction, Types, characteristics & Examples.pdf. 1-8.

Howard, J.H.; Kazar, M.L.; Menees, S.G.; Nichols, D.A.; Satyanarayanan, M.; Sidebotham, R.N.; West, M.J. (1988). Scale and performance in a distributed filesystem. ACM Trans ComputSyst (TOCS*)* 6(1). 51–81.

IDC. (2011). The 2011 Digital Universe Study: Extracting Value from Chaos. [Online] Available from: http://www.emc.com/collateral/analyst-reports/idc-extracting-value-from-chaos-ar.pdf [Accessed 9th July 2012].

Laney, D. (2001). 3D data management: Controlling data volume, velocity, and variety. In: Meta Group. Available at: http://blogs.gartner.com/doug-laney/files/2012/01/ad949-3D-Data-Management-Controlling-Data-Volume-Velocity-and-Variety.pdf (accessed 16 January 2013).

Rijmenam, M. V. (2014). Think Bigger: Developing a Successful Big Data Strategy for your Business. Available at *Amazon.*

Shukla, Soumya et al. (2015). Big Data: Concept, Handling and Challenges: An Overview. *International Journal of Computer Applications.* 114 (11). 6-9.

Soon, KWK et al. (2016). A study of the determinants affecting adoption of Big Data using integrated Technology Acceptance Model (TAM) and diffusion of innovation (DOI) in Malaysia. Pdf (researchgate.net)

9

Radio Frequency Identification

Radio Frequency Identification (RFID) is a touch-free technology offering increased accuracy and real-time inventory tracking at low cost. RFID uses radio waves to read and capture the information. The technology helps in storing large amount of information on chips (tags/transponders) which can be read at a distance by a reader without requiring the line-of-sight scanning. Figure 9.1 shows a typical RFID system. RFID readers scan tags or transponders placed on an item, for example, a pallet with many boxes, a box having many items, or a single item. The readers pass on the information received from tags to the backend system which typically comprises of a database and the RFID application. The backend application communicates the information received from tags and readers to the Supply Chain partners.

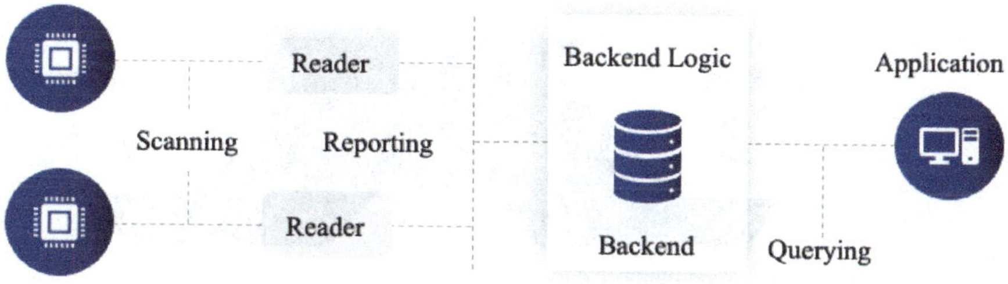

Figure 9.1: Typical RFID system

9.1 Evolution of RFID technology

RFID was first time used during World War II in conjunction of radar technology to find out if the airplane is of enemy or our own side. During 1970s, Electronic Article Surveillance (EAS) was used to detect thefts in retails. Any unscanned item passing through the exit doors will raise an alarm and the theft will be noticed (Miles et al., 2008). RFID was used in tracking the movement of the animals, industrial use, and payment of toll roads during the 1980s (Rochel, 2008). Nash (2010) reports use of RFID in many applications including inventory tracking, animal tracking, passport recognition, country/state ID cards, food tracking. The technology is supporting industry in lowering waste, theft, inventory costs, and providing a more accurate inventory count. The possibilities are endless with RFID technology.

Demand planning capabilities are strengthened with the help of more reliable and up-to-date information thus potentially reducing out-of-stock and overstock situations with multiple options in procurement.

9.2 Components of RFID

RFID system consists of four components: Tags (Transponders), Readers, Software, and Network (Nash, 2010) (see Figure 9.2 below).

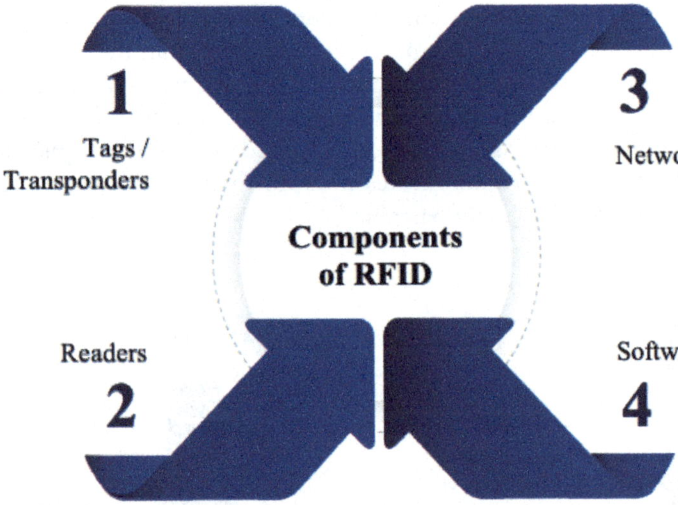

Figure 9.2: Components of RFID system

9.2.1 Tags

Tags come in three types: passive, active, and two-way (semi-passive). Feature of the three types of tags is given in Table 9.1. Tags can be used only once or multiple times. Single time use tag is attached to a single item purchase. After the use, tag is thrown to waste. A multiple-time tag is fixed on the window of the truck enabling it to enter the plant or warehouse on a reusable pallet holding the multiple products on shipment.

9.2.2 Readers

A reader can be hand-held or fixed type. Hand-held readers can read or write data everywhere and are mobile. Fixed readers are installed at the main gates to identify people, goods, or vehicles entering the place. Fixed type readers can also be fixed in a shelf to transmit inventory information (Wisner et al., 2009).

TABLE 9.1: FEATURES OF DIFFERENT TYPES OF TAGS

No.	Type of Tag	Features of the Tag
1	Passive type	Read only, smaller, least cost, provide shorter range, enhanced durability, and do not have a battery.
2	Active tags	Read/write, large memory, and longer range, more costly, bulky, limited life, have a battery
3	Two-way (semi-passive)	Tags can communicate in and out without reference to an antenna

9.2.3 RFID Software

There are four main layers that make up the software. The first layer consists of the management of the devices with different configuration parameters and control actions that must be executed for optimal operations. The second layer is used for data acquisition from the RFID reader. The third layer has to do with actions that have to do with the data itself. Data has a half-life making it necessary for workflows to be performed in real time. The fourth and highest layer of the software is interfacing, where information systems enable companies to interface with each other and share data about goods tagged with EPC tags (Whitaker et al., 2007).

9.2.4 Communication Network

A communication network is needed for companies to interact with each other and help in resolving issues with security and authentication between known trading partners. It can also help further develop the relationship between the trading partners enabling more trust and reliability.

9.3 Application of RFID

RFID has been adopted by different industry sectors in different applications. Some of the applications are discussed in Table 9.2.

TABLE 9.2: APPLICATIONS OF RIFD IN DIFFERENT INDUSTRIES

No.	Industry	Activities to use RFID
1	Retails	• Track the inventory from the supplier to the point of sale. Example: Walmart, USA
2	Manufacturer	• Track the items produced by giving full information on the tag. • A car manufacturer recalled defective cars using RFID to track the problem and fix it.
3	Pharmaceuticals	• Track bottles of drug with higher cost having chances of theft.
4	Hospitals	• Tracking the drug in short supply, reducing waste and theft. Shortage of Corona vaccine is a good example that explains the need for tracking. • Advanced hospitals release a certain number of items for regular care, but there is also a device in store for emergency circumstances.
5	Government	• RFID technology is used for identification, security, and defense purposes.
6	Passport Office	• China, India, USA, and the UK attach RFID tags to Passport.
7	Defense Agencies	• RFID is used to trace their military supplier's activities and verify that the products are made correctly. • High security areas can also install the security at entrances to verify an approved person or vehicle.
8	Airline Industry	• To track their customer's checked luggage. If a bag is lost or stolen, the airline can locate the baggage with this technology. • It is easier to reroute luggage if the flight is changed due to the airline or customer.
9	Banks	• The tags can be placed in monetary documents such as cash or bonds to secure that it is real.
10	Food Industry	• To track meat, poultry, and produce to ensure the safety of the item, beginning to end.
11	Logistics and transportation	• To track the movement of goods throughout the network, including the returns. • Shipping companies track movement of containers. • Tracking radioactive and explosive materials is an important example.

9.3.1 Impact of RFID in Retail Sector

RFID has made an impressive impact on various processes in receiving, backroom inventory management, store floor, and promotions and sales in the retail sector. Figure 9.3 presents benefits achieved with the introduction of RFID in retail in receiving, store floor, backroom management, and promotions and sales (Charikleia, 2018).

9.4 Challenges in RFID Implementation

RFID technology has been very versatile and has been found very useful. However, its implementation has not always been easy and smooth. Some of the challenges in implementation of RFID are:

i. A lack of management and supply chain partner commitment. This is the biggest challenge that companies face in adoption of RFID.

ii. High cost of the technology comes as another issue as a hindrance.

iii. Legal and patent problems are at times very time consuming and irritant in the implementation of the technology.

iv. Operational automation finds problems.

v. High cost of the technology comes as another issue as a hindrance.

vi. Legal and patent problems are at times very time consuming and irritant in the implementation of the technology.

vii. Selection of hardware and software.

viii. Technology support for adoption lacks in some cases.

9.5.2 Benefits of RFID to Walmart

Implementation of RFID reduced the problem out-of-stock situations by 8 percent of stock items benefitting the retailers and their suppliers across the world (Collins, 2005a). Gillette and Walmart used RFID improving sales by 19 percent through promotional efforts. Promotional items were kept in distribution centers, the back-stores, and promotional displays offering items on-time and avoiding stock-outs (Kaisri et al., 2012).

Bhuptani and Moradpour (2005) report following advantages from the implementation of RFID by Walmart:

i. Reduction in data entry errors

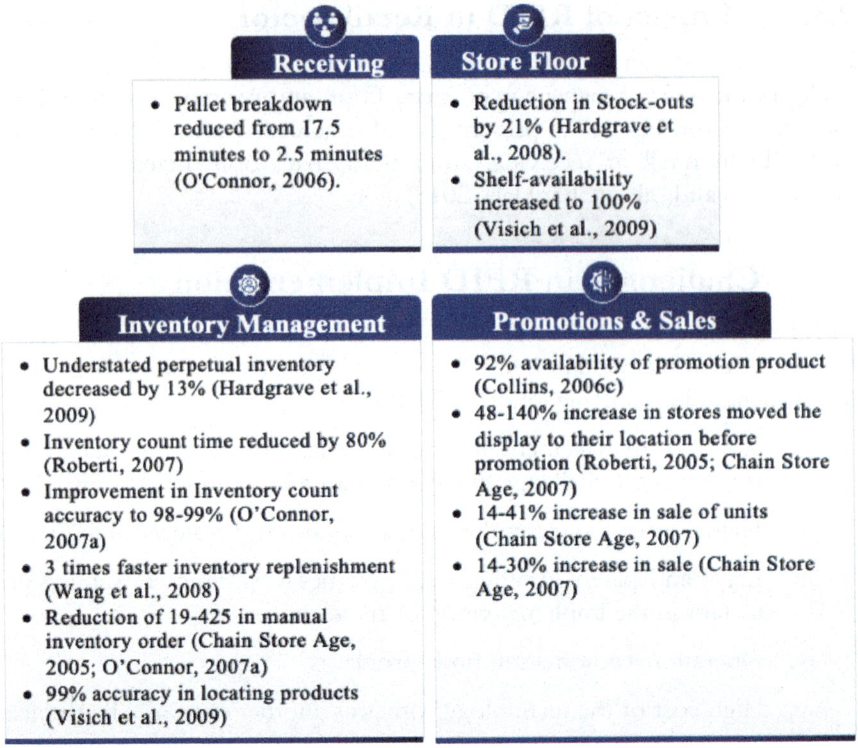

Source: Charikleia (2018)

Figure 9.3: Impact of RFID in Retail Sector

ii. Automation in receiving, inventory management, shipping, and payables replacing the manual processes.

iii. Automatic tracking product movement leading to reduction in the size of the inventory and out-of-stock situations.

iv. Customer's satisfaction improved

v. Availability of the right product at the right place, and at right time

vi. Easy demand forecasting for the products

vii. Shorter time in shopping

Hardgrave et al. (2008) studied 4554 items in 24 stores of Walmart and observed that stock-outs reduced by 26 percent at 12 stores where RFID has been implemented.

9.6 Summary

RFID is extensively used by industry in various supply chain activities. This Chapter explains a typical RFID system. The Chapter traces the evolution of the RFID technology and discusses various aspects of the technology. RFID systems have four components: Tags, reader, software, and network. The RFID technology has wide-spread application in lowering waste, theft, inventory costs, and provides more accurate count. The technology helps in strengthening the demand planning capabilities and procurement. The Chapter tracks various applications in different sectors like retail, manufacturing, pharmaceuticals, hospitals, logistics, and transportation. Case of Walmart provides insight of benefits from implementation of RFID in the case of retail, Various challenges faced by the RFID technology have been covered in detail.

References

Chain Store Age. (2005). Wal-Mart Quantifies RFID Benefits. Retrieved April 04, 2017, from http://business.highbeam.com/410731/article-1G1-140216817/researchconfirms-walmart-rfid-benefit

Chain Store Age. (2007). Best Buy's best-cost, customer-centric supply chain. Retrieved April 04, 2017, from http://chainstoreage.com/article/best-buy%E2%80%99s-bestcost-customer-centric-supply-chain-0

Collins, J. (2006c). RFID implementation is an art. Retrieved April 04, 2017, from http://www.rfidjournal.com/article/articleprint/2427/-1/1/

Hardgrave, B., Aloysius, J., & Goyal, S. (2009). Does RFID improve inventory accuracy? A preliminary analysis. International Journal of RF Technologies: Research and Applications. 1(1), 44–56. doi:10.1080/17545730802338333

Miles, Stephen B, Sanjay E Sarma, and John R Williams. (2008). RFID Technology and Applications. Cambridge, New York: Cambridge University Press. 1-244.

Nash, Tina A. (2010). RFID Technology and its Impact on the Supply Chain. Senior Honors Thesis. 243. Eastern Michigan University. 1-50.

O'Connor, M. C. (2006). Staples business depot sees big benefits from RFID test. RFID Journal. Retrieved April 17, 2017, from http://www.rfidjournal.com/article/articleprint/2684/-1/1/

O'Connor, M. C. (2007). Falabella plans second item-level RFID pilot. Retrieved April 11, 2017, from http://www.rfidjournal.com/article/articleprint/3585/-1/1/

Roberti, M. (2005). The serendipity effect. Retrieved April 13, 2017, from http://www.rfidjournal.com/article/articleprint/1960/-1/128/

Roberti, M. (2007). RFID heats up in Latin America. Retrieved April 13, 2017, from http://www.rfidjournal.com/article/articleprint/3344/-1/2/

Rochel, Roman. (2008). RFID Technology and Impacts on Supply Chain Management Systems. Saarbruken, Germany: VDM Verlag Dr. Muller Aktiengesellschaft & Co. Saarbrucken, Germany. 1-96.

Visich, J. K., Li, S., Khumawala, B. M., & Reyes, P. M. (2009). Empirical evidence of RFID impacts on supply chain performance. International Journal of Operations & Production Management, 29(12), 1290–1315. doi:10.1108/01443570911006009

Wal-Mart. (2016). Annual report. Retrieved April 20, 2017, from https://s2.q4cdn.com/056532643/files/doc_financials/2016/annual/2016-Annual-Report-PDF.pdf

Wang, L., Gwebu, K. L., Wang, J., & Zhu, D. X. (2008). The aftermath of information technology outsourcing: An empirical study of firm performance following outsourcing decisions. Journal of Information Systems, 22(1), 125–159.

Whitaker, Jonathan, Sunil Mithas and M. S. Krishnan. (2007). A field study of RFID deployment and return expectations. Production and Operations Management Society. 599-612.

Wisner, Joel D, Keah-Choon Tan and G. Keong Leong. (2019). Principles of Supply Chain Management: A Balanced Approach. Mason, OH: South-Western Cengage Learning, 209.

10

Simulation Technology

Businesses operating in a competitive environment are in search of effective methods to reduce costs with fast and correct decisions in a pressured, competitive, and fast changing market environment. Supply Chain Management (SCM) offers great opportunities enabling businesses to design and optimize the supply chains. Simulation models for SCs are expected to improve the efficiency of SCM and reduce the cost. Simulation-based optimization techniques are in great demand in simulation for choosing the best design. Simulation technology reflects the physical world in a virtual model that enables optimisation of the machine or service system settings before implementing physical changeover. Users can find out optimal operating conditions by virtually examining an operation using different scenarios through simulation modeling.

General-purpose simulation package has a wide range of applications such as transportation system design, logistics support analysis, supply chain optimization, customer service, internal business processes, and six sigma projects etc. The simulation packages are designed for different special functions or enterprises based on their needs. Industries like manufacturing, energy, nuclear, aviation, food & beverage, defense, financial services, call centers, and government use simulation packages. The software packages are designed for easy-to-use, quick modeling, and are extremely interactive and visually real.

10.1 Simulation Applications in SCM

Supply Chain deals with a variety of entities such as plant and machinery, manufacturing plants, and people engaged in information sharing, collaboration, and a number of other activities. Simulation of the scenario due to these factors face new and greater challenges. Zheng et al. (2008) cite a number of examples of use of simulation technology in improving performance in different areas and industries. A survey (Marco et al., 2006) conducted in the areas of semiconductors and automobile industries in the USA focused on the simulation applications. A fluid simulation model using Petri Net was made to study many events in the Distribution Center (Ying and Chen, 2005). ARENA and EXCEL simulation models were used to study the product distribution in the supply chain (Lee et al., 2006). Study of relationship between information sharing and SC configuration used a parametric simulation model using Arena software to evaluate various supply chains (Rong et al., 2006). Simulation techniques employing Smart SCOR model have been used by the managers in making decisions in improving business processes, and inventory allocation (Hongwai et al., 2006). Extend software was used in simulation of the real-time output analysis for a steel plant (Peter et al., 2000). It is reported that IBM saved US Dollar 100 million in inventory management with the use of ATS simulation method (Thomas et al., 2006).

10.2 Dynamic Simulation

Supply chain decisions are based on details available to the managers about the problems. anyLogistix has classified the problems in three categories with their level of abstraction and the level of details available. Table 10.1 shows the three levels and the types of the problems related to these levels. The quality of decision-making depends upon these levels. Poor decisions will be linked to the problems where more abstraction and less details are available whereas better quality decisions are based on less of abstraction and more of details available.

Dynamic simulation is being adopted by some of the companies to empower themselves with wide a range of capabilities. The model 'dynamic simulation' details out the system and the operating rules for simulating the business process logic and the interdependencies of the system components. According to anyLogistix, the developer of the model and the software, the system is dynamic because one can run the system and can observe how the system operates over time in the same way as one plays computer games. The system has two digital components of the supply chain. As the first component, it permits someone to experiment with improvements and the second component is that the user can see the benefits. The advanced version of the system has inbuilt flexible logic that can be adjusted and customized to the customer's requirements.

Till recently, spreadsheet-based modeling and calculations have been commonly used for solving the high abstraction level business issues. Spreadsheets can solve simple problems of the supply chain, but they have been found incapable of solving the complex issues. In case of complex problems, industry all over the globe has adopted analytical optimization and dynamic simulation.

TABLE 10.1: TYPES OF PROBLEMS & LEVELS OF ABSTRACTION / DETAILS

No.	Abstraction	Types of Problems
1	High abstraction level – low detail	Network-level problems, determined by basic supply chain structure and dependencies.
2	Medium abstraction level – more detail	The details of supply chain organization, considering policy parameters, resources, process logic, randomness, and operational mics.
3	Low abstraction level – high detail	Determining how processes inside the organization impacts the whole supply chain considering the processes, resources, and constraints.

Source: anyLogistix (nd)

Analytical models suit very well to solve supply chain challenges at the network level, where operational logic, randomness, and dynamics are not of major concern. In cases, where considering any of the factors like logic, randomness, and dynamics is of importance, the analytical optimization can be used to get first approximate solution which should then be followed with detailed analysis with dynamic simulation modeling (anyLogistix, nd). Integrating the analytical optimization and dynamic simulation offers great opportunities in cost reduction and increased customer satisfaction (anyLogistix, nd).

10.3 Dynamic Simulation in SCM

Supply chain management faces several challenges, which can be managed well with the use of dynamic simulation. Table 10.2 presents the challenges usually faced by the supply chain and the problem solutions offered by dynamic simulation.

In simulation modeling the attempt is not on optimization but certain scenarios are simulated using the model and the result is seen. Model results are observed in detail and understand how the supply chain changes over time. Number of simulation efforts are tried and one with the best result is selected for a management decision.

Dynamic simulation enables the user to see the results and also be able to see how the result has arrived. Impact of each event can be traced on the supply chain. Conclusions based on the simulations are real time data rather than on guess work. It is possible to measure at the same time various metrics in dynamic simulation. Examples are costs, Distribution Centre capacity utilization, service levels.

TABLE 10.2: HOW DYNAMIC SIMULATION ADDRESSES TYPICAL SCM CHALLENGES?

Challenges	How Dynamic Simulation Helps?
New supply chain design implementation	To introduce a solution offered by network optimization. **Example**: How to implement a solution suggested by the optimization, which is unimplementable?
Supply Chain operations	Understanding the operations considering the resources and the policy in detail over the time for efficient management of the network. **Example**: Working out safety stock for each product at each facility.
Supply chain innovation trials	How to improve operational performance of new high-level network design? **Example**: Testing the innovative options before implementation.
Risk assessment	Assessing the risks related to the proposed supply chain structure.
Process assessment within the organization	Assessing the impact of the internal processes on the operations in the distribution centers or the factories in the entire supply chain.

10.4 Benefits of Simulation Technology

Simulation technology offers benefits to the parties in the supply chain network in many ways. Kelton et al. (2004) discuss some examples that are presented in the following sub-chapters.

i. **Impact of Unexpected Events**: Users may model unexpected or unforeseen events in certain areas such as impact of an earthquake, a severe storm, sudden flooding of an area, change in the government in a country, spread of a wildfire, and understanding the impact of these to the various enterprises in the SCs.

ii. **Change in Plans**: Users may consider multiple options to mitigate risks in advance before executing specific planned changes, for example: To assess the effect of some innovative changes to be introduced in the existing infrastructure or adding a new one within the SC; strategic operational changes, such as process, location, and use of new facilities; the fusion or the separations of some components of two SC; manufacturing products inside the company;

and creating new suppliers or subcontracting some processes, the simulation technology can support the decision with useful analysis.

- In uncertain and unpredictable situations like the surge of COVID-19 all over the world, companies must change or rationalize the number and size of order lots, the total costs, quality, flexibility, and the responsibilities. The changes may be in one or more parameters at the same time. Simulation technology provides the best solution for solving the problems in the supply chains and would help in maintaining good relationship with the suppliers satisfying the interests of both sides.
- Simulation is an extremely useful technique to investigate the opportunities to reduce the varieties of product components and standardize them throughout the Supply Chains.

10.5 Future of Simulation Technology

Simulation in the supply chain has a very promising future due to its capabilities in providing the information support in decision making. There is potential in extending the applications of simulation in the supply chain. Simulation is not limited to design the supply chain but provides wide range of information to users in running the system also. Simulation applications could be extended to PLCs and enables engineers to understand the running status of the equipment before installation. The business process planning can be integrated with the extended application of simulation so that managers and engineers can know the status of their equipment. This will reduce errors in manual entry of data as the data is integrated. A new trend was started in integrating different models to create another simulation model. Different enterprises of the supply chain can integrate their own software with different commercial off-the-shelf simulation packages. There are many other similar integration examples in supply chain applications. All these opportunities provide a wide scope of growth opportunities to simulation in the supply chains.

10.6 Simulation Case Study

The telecom company with a nation-wide transportation network offers Business-to-Business (B2B), Business-to-Customer (B2C), and Business-to-Government (B2G) telecom services. The company operates a very large single-tier logistics network with about 150 warehouses, 2,500 points of sales, and about 20,000 operational sites (Deloitte CIS, 2017). The company hired the services of Deloitte CIS to optimize its widespread telecom network and develop an optimal model for demand forecasting, optimizing the inventory and maintaining efficient services at operational sites and points of sale.

The telecom giant set the following objectives:

i. Assessing the optimal hierarchy for the network – 1 or 2-tiers.

ii. To set the costs and effects of implementing various inventory optimization policies including optimization of the network.

iii. To decide the capacity of the warehouses including their locations to meet the future demand and inventory optimization targets.

iv. To develop an optimal network transformation plan, inventory relocation plan, and setting the timeline for closure of warehouses.

Deloitte CIS (2017) obtained the telecom company's current logistics network data from the company's ERP system and created a baseline model with approximately 99% accuracy. Approval from key stakeholders was obtained for the baseline model. Deloitte CIS used anyLogistix software to design six what-if scenarios with different supply chain structures and inventory policies. anyLogistix is a new optimization tool that combines the classic supply chain design capabilities with dynamic simulation in one platform. The Deloitte CIS team then ran optimizations on each scenario and found the best option for the telecom company based on cost, service level, and implementation risks. Deloitte CIS created visual dashboard reports for each scenario to get validation of results with stakeholders.

Using anyLogistix, Deloitte consultants developed a step-by-step roadmap of transition from the current supply chain design to the target state. Recommendations on inventory policies for different supply chain optimization scenarios were developed. The optimization project led to significant increase in logistical efficiency by using the simulation software anyLogistix, the details are given as detailed below:

i. Reduction in total logistics costs by 9%.

ii. Reduction in number of warehouses by 28%

iii. Reduction in transportation costs by 5%

iv. Improvement in inventory turnover by 26%

v. Reduction in total inventory level by 18%

10.7 Summary

Simulation models for supply chain are expected to improve efficiency of SCM and reduce the cost. Simulation models are in great demand for selecting the best design and are becoming essential element in cases of high-cost investments and risky projects. Simulation studies address this problem by providing advance information about the likely success or failure. Investors prefer to know in advance the expected returns for their investments along with a timeframe for returns from the investment. The chapter discusses various applications of the simulation studies in different sectors such as manufacturing, energy, nuclear, aviation, food & beverage, defense, and financial services.

Simulation applications in SCM have found their place in many areas including improvements in operational performance. Distribution centers has extensively used simulation modeling. Simulation has been used for quick decision making based on real-time information. The Chapter presents problem types and levels of abstraction with the help of dynamic simulation. Benefits of simulation technology are presented in detail for different areas. The case study of the Russian Telecom Company shows the extent of benefits obtained from optimization studies using simulation technology.

References

anyLogistix. (nd). Supply Chain Optimization and Simulation: Technology Overview. *AnyLogic Company*. North America. 1-23.

Deloitte CIS (2017). Supply Chain Network Design for a Telecom Provider. pdf presentation. https://www.anylogistix.com/supply-chain-network-design-for-a-telecom-provider/

Hongwei, D., Changrui, R., Wei, W., Jin, D. (2006). Applying simulation in a supply chain transformation case. The 38th conf on Winter simulation, Monterey, California.

Kelton, W.D., Sadowski, R.P., Sadowski, D.A. (2004). Simulation with Arena, Second Edition. McGraw Hill. 1-700.

Lee, Y. K. et al. (2006). Evaluating refinery supply chain policies and investment decisions through simulation-optimization. The 38th conf on Winter simulation, Monterey, California.

Marco, S., Hakon, F., Jan Ola, S. (2006). Applications of discrete-event simulation to support manufacturing logistics decision-making: a survey. The 38th conf on Winter simulation, Monterey, California.

Peter, M., Srinagesh, G., Douglas, J.M. (2000). Strategies for successful simulation of manufacturing systems: dynamic output analysis for simulations of manufacturing environments. The 32nd conf on Winter simulation, Orlando, Florida.

Rong, L., Akhil, K., Alan, J.S. (2006). Simulation results for supply chain configurations based on information sharing. The 38th conf on Winter simulation, Monterey, California.

Thomas R.E. et al. (2006) Simulating order fulfillment with product substitutions in an assemble-to-order SC. The 38th conf on Winter simulation, Monterey, California

Ying, W., Chen, Z. (2005). Fluid based simulation model for high volume DC conveyor systems. The 37th conf on Winter simulation, Orlando, Florida.

Zheng, X., Yu, H.; Atkins, A.S. (2008). An Overview of Simulation in Supply Chains. In book 'Advanced Design and Manufacture to Gain a Competitive Edge. 407-416. ResearchGate.

11

Software in Digital Supply Chain Management

Efficient and effective supply chain provides all the input materials and other supporting spare parts and consumable to ensure uninterrupted production of goods. Supply should be right in terms quantity, quality, at right time (just-in-time), at right place, and at right price. Right selection of source of supply is of great importance to satisfy all the rights. The company wants its suppliers to be capable of supplying good quality materials in required quantity, defined quality, at the most optimized cost, as per given schedule, and at the right place. Maintaining good relations with the supplier is as important as the relationship with the customers. A company needs to have good and collaborative relationship with its suppliers and must satisfy its customers by providing good materials on time in specified quantity and quality. Since the company deals with hundreds of suppliers and thousands of customers, it needs strong tools in form of good software to empower the company with dependable support. Users of the software should find the suppliers of the software technically competent and financially resourceful.

11.1 Supply Chain Management Software

SCM Software is a real time management software that supports all the activities in the supply chain. The software manages the demand forecasting. demand planning, supplier sourcing, request for quote (RFQ), order placing, production planning, delivery schedule for raw materials, production of the finished product, storage, delivery to the distribution centers, delivery to ultimate customer, warehouse management, inventory control.

The software manages the flow of the product, information, and the flow of money. Software connects the suppliers with the enterprise on one hand and the enterprise with the customers and provides real-time information. A multi- national company (MNC) and global company may be having its manufacturing facilities and distribution centers in a number of countries. The SCM software connects everything in the network, supporting the concept of Internet of Things (IoT).

The objective of the SCM software is to optimize and make every activity in the supply chain efficient, economical with delivery of quality product or quality services. Today, SCM software does not stand on one technology but involves a number of technologies. Selection of right software and right technologies depend upon the needs of the enterprise. As the needs of each enterprise differs due to its size, production process, raw materials, type of labor force, and types of customers, the choice for SCM software will also differ.

The SCM software covers all the processes from the stage of product development to the logistics, warehousing, inventory control, and transportation. The functions of the software do not stop here but go into performance evaluation of each supplier, performance of each section or department of the enterprise. The software also helps in evaluating customer satisfaction. The enterprise becomes more efficient, competitive, and economical with good revenues with the use of the software.

11.1.1 Features of SCM Software

Supply Chain Management software has all the features that cover all the business activities. The activities in the supply chain include demand forecasting, inventory management, order planning, request for quote (RFQ), supplier sourcing, analytics of offers, order placing, production and delivery of products, transportation of goods, store keeping and warehousing management, payment to supplier, reverse logistics, and performance evaluation. All the materials such as raw materials, spare parts, components, assemblies, finished goods, and reject materials are covered by the software. Quick payment to the suppliers helps the purchasing group to negotiate good prices and favorable terms and conditions and in developing a relationship collaborating with the supplier.

11.1.2 Benefits of Typical SCM Software

Supply chain management software offers number benefits to the enterprise (PAT Research, 2021), such as:

i. **Efficient Operations**: Efficient operations helping to deliver the finished products to the customer on scheduled date.

ii. **Cost Optimization**: Supply chain management software helps in reducing both the fixed as well as variable costs and lead to cost optimization.

iii. **Reduction in Processing Time**: The software reduces problems due to better communication and collaboration such as delayed product shipment from suppliers, hold-up in the chain of production, and logistics errors.

iv. **Higher Profits**: The software enhances the visibility and collaboration that leads to higher production and thereby higher profits.

v. **Lowering the Liability and Risks**: Software informed the concerned parties in advance about the likely risks, who take corrective action.

vi. **Improved Customer Service**: Software improves the shipping and delivery of the items thus improves the customer's service.

vii. **Better Relationship with Suppliers**: The enterprise software can analyze various parameters and sort out issues, if any. It is easy to review: What rates are offered by the supplier in comparison to others? What has been the delivery service of the supplier in comparison to others? Information on such issues drives a healthier relationship between the enterprise and the suppliers.

11.2 Top SCM Software

PAT Research, a B2B discovery platform, is engaged in research, analysis for Enterprise Software and Services, and many other activities. Based on its research, PAT Research has suggested a list of top15 software in 2021 that includes E2open, SAP SCM, Logility, Perfect Commerce, Oracle SCM, Infor SCM, JDA SCM, Manhattan SCM, Epicor SCM, Dassault Systems SCM, Descartes SCM, HighJump SCM, IFS, Watson Supply Chain, BluJay SCM for supply chain management (PAT Research, 2021). This list is not the end-list as new software are added from time to time e.g., Robotic Process Automation (RPA) software and anyLogistixTM are software's, developed using a combination of digital technologies for the supply chain management. This Chapter briefly introduces the top 15 software in the year 2021, identified by PAT Research (2021), and the RPA, an important development in the field using more than one digital technologies.

11.2.1 E2open

E2open is an online supply chain management system designed to offer a more responsive and demand driven supply chain (PAT Research, 2021).

Features: The software supports all activities of supply chain management starting from supply chain planning, analytics, and monitoring. It also covers order processing, sourcing, procurement, supplier management including collaboration with the client. In addition, it also covers inventory management, warehousing management, and transportation management. The software uses cloud technology and simulation in collaborating internal operations of the organization with the external operations of partners and clients.

11.2.2 SAP SCM

The software is offered by SAP, and leverages AI and IoT technologies to help customers gain visibility and transparency of their supply chain using advanced analytics features and functionality (PAT Research, 2021).

SAP SCM offers different software such as SAP integrated business planning, SAP transportation management, SAP extended warehouse management, supply chain collaboration and strategic sourcing (PAT Research, 2021).

Features: The software integrates the demand forecasting, inventory planning, supply planning, and financial planning. The software also covers inventory management, stores management, and transportation management. SAP SCM enables collaboration with partners, suppliers, and clients. It offers digitization and automation from source to contract process for all the spend categories. Small, medium, and large size customers can be served with this software.

11.2.3 Logility

Logility offers its software named as 'Logility', a supply chain management software, with features of advanced analytics and optimization of planning that empower the organization to act with the market proactively (PAT Research, 2021).

Features: The software can integrate collaborative scenario analytics and advanced analytics. The software integrates optimization of demand, optimization of inventory, management of data, and transportation management. The supply optimization supports simultaneous sourcing, multi-tiered distribution, and multi-sourcing capabilities. The supplier analysis feature of the software gives visibility that helps in cost reduction, and strategies to efficient transportation and freight.

11.2.4 Perfect Commerce

Perfect Commerce offers its software named as 'Perfect Commerce', a supply chain management software, with ability to suggest multiple solutions for maximization of 'Return-on-Investment' (ROI) and simplify the process of source-to-pay.

Features: The software offers implementation services, management of tail spends, service for auctions, catalog content services, group purchasing, and strategic sourcing (PAT Research, 2021).

Perfect Commerce helps in bringing together the subject matter experts and the stakeholders to develop specifications, document specification, line-of-action for the project, requirements, and outcomes. The facility of group purchasing helps in getting maximum discounts that lead to a reduction in cost and an improvement in productivity.

11.2.5 Oracle SCM

Oracle's Supply Chain Management software changes the business of the users, with end-to-end visibility starting from the product development to fulfilling orders through planning and inventory management, sourcing, manufacturing, and delivery (PAT Research, 2021).

Features: Oracle's supply chain management uses cloud technology in planning, order management, and production. With the use of these features, users can develop Cloud proactive devices revenue growth and strategies for the reduction in cost. Using Cloud technology, even inexperienced users can use visual analytics and intelligent recommendations. Manufacturing clouds stream lights users manufacturing processes. Oracle Planning Central Cloud is a modern and easy-to-use supply chain solution.

11.2.6 Infor SCM

Infor SCM offers its software named 'Infor SCM' to integrate advanced supply chain planning, finance capabilities, and implementation with the GT Nexus Commerce Network to help companies to compete in the market with proper control and speed (PAT Research, 2021). The software is found suitable to meet the requirements of small and medium size enterprises.

Features: Supply chain visibility, planning of the supply chain, warehouse management, collaboration with suppliers, supply chain finance, operations management, and transportation management are some of the features of the software. The enterprise using the software has visibility to find weak spots and manage them well to avoid potential disruption in supplies including taking care of demand signals. The visibility offered by the software enables the enterprise to minimize the safety stocks of the inventory and the risks. The software helps in efficient operations within the supply network.

Another important feature of the software is its capabilities in helping the suppliers of the company and the customers to collaborate in forecasting, commitments, and plans and enable the company to respond to the dynamic and changing demand. Transportation management module of the software enables movement of freight from source to destination efficiently, reliably, and cost-effectively.

The supply chain finance feature helps in driving accurate and timely payments to suppliers, as well as receiving/collecting payments from customers as quickly as possible.

11.2.7 JDA SCM

JDA SCM offers supply chain management software named 'JDA SCM'. The software supports end-to-end retail and supply chain planning and execution using Cloud technology and meets the needs of the present global and consumer-driven marketplace that are dealing with complex issues and situations (PAT Research, 2021). The software has been found suitable for small and medium size organizations.

Features: JDA SCM software helps the companies to effectively manage manufacturing and distribution, retail and service industries with solutions supported with Cloud technology. The Cloud based solutions enable the users in saving costs, increasing profitability, and offering enhanced collaboration right from planning stage to the delivery stage. Ensuring a committed delivery schedule is one of the key features of the software. Using the JDA SCM's manufacturing and distribution management solution, users can collaborate with the retailers to create localized inventory solutions at store-level.

The manufacturing and distribution Category Management solution enables analytics-driven knowledge of the shoppers' likes and dislikes at the macro as well as micro and local level markets. Users can prepare localized and actionable plans for the profitable growth. Users can reduce the inventory levels and save costs. The customer service gets improved with quick responses.

JDA SCM's Retail Solutions provide real-time inventory updates in managing stores combining the budgeting, forecasting, Warehousing, and inventory management helping the users to enhance stores operations and optimize on costs. Companies in the hospitality, travel, transportation, and media industry have improved their services and profitability with the use of this software.

11.2.8 Manhattan SCM

Manhattan SCM has named its software as 'Manhattan SCM' that offers designed or worked out solutions demolishing the departmental / organizational boundaries and serving the needs of the supply chain (PAT Research, 2021). The software is found to be useful for small and medium sized organizations,

Features: The software provides solutions in the limited areas relating to carrier management, distribution of goods and supplies, and transportation overview. The carrier feature of the software serves well in assignment loading, reducing the fuel costs, effective utilization of drivers and the vehicles. The most important contribution of this software is that it serves well in fulfilling customer's requirements. The software is extremely useful in situations of varying and unpredictable nature such as in the period of Corona-19.

The features on distribution and transportation refer to the management of warehouse, billing, labor, and space utilization. The success of e-business rests on these areas of operations.

Business solutions offered by the software save transportation costs, enhance the level of services, and reduce the operational risks.

11.2.9 Epicor SCM

Epicor SCM has named its software as 'Epicor SCM' which provides a platform for improvement in the efficiency of the supply chain. The software uses Cloud technology for improving the operational parameters (PAT Research, 2021).

Features: The software caters to demand management, purchase management, inventory management, warehouse management, advanced materials management. The software helps in managing supplier relationships and connecting with the suppliers. The software works with cloud technology and empowers the user with centralized solutions that provide data visibility, and in depth understanding. Inventory management module of the software helps in automation, digitization, and simplification of processes.

11.2.10 Descartes SCM

Descartes SCM software of Descartes SCM company is supply chain management software that uses Cloud technology, modular, wireless logistics, and interoperable web solutions (PAT Research, 2021). The software is suitable for small, medium, and large companies.

Features: The software has a limited coverage in comparison to other software. It takes care of routing, transportation management, customs, and regulatory compliance. In addition, the software also covers the activities of brokers and forwarders. The transportation management feature provides increased operational efficiency, lower costs, less complexities, and better customer service.

The broker and forwarder enterprise systems help in increasing productivity by optimizing the freight, and compliance with security and fiscal obligations.

11.2.11 Dassault Systems SCM

Dassault Systems SCM has named its software as Dassault Systems SCM is dedicated to automotive suppliers and manufacturers. The software offers complete sets of tools for managing the sourcing cycles (PAT Research, 2021). The software is extremely useful to small and medium size companies.

Features: Dassault Systems SCM software offers a platform to prepare the Request for Quotations. The focus of the software is on procurement, contracting and collaboration with suppliers. The approach in procurement takes into consideration the whole life cycle. Supplier database management, contract management, concurrent design management, component lifecycle management, reuse and usage tracking and supplier collaboration are the main features of the software.

11.2.12 HighJump SCM

HighJump SCM has named their software as 'HighJump SCM'. The software offers a platform to respond to market needs using the supply chain technologies (PAT Research, 2021). The software meets the needs of small, medium, and large-scale enterprises.

Features: Management of labor, multi-client operations, and integration with ERP are some of the special attractions of the software. The yard management feature of the software enables tackling complex issues. Intelligent work direction, real-time information, and full visibility of the vehicles, drivers, and the goods are covered well by the yard management feature. The software is an extremely useful tool in cost reduction, and in improving productivity of the enterprise. Flexible management is especially suitable with changing scenarios and uncertain events.

11.2.13 IFS SCM

IFS has named its supply chain management software as 'IFS'. The software offers simple visualization of the flow of products and adds to the efficiency of the whole supply chain management system (PAT Research, 2021). IFS is suitable for small, medium, and large size companies.

Features: The main features of the IFS include demand forecasting, purchasing, management of rental, and warehousing management. The software offers highly graphical and interactive tool to conduct highly effective forecasting.

It is easy to prepare a master production plan, materials requirements with delivery schedule, distribution orders, and calculate required capacity with the help of IFS software.

The custom order feature is extremely useful for multinational and global companies to operate with centralized procurement and make good savings through economic purchasing. The warehousing feature covers operations right from goods receipt to point of issue. The software has many other attractive features.

11.2.14 Watson Supply Chain

Watson Supply Chain has named its software as 'Watson Supply Chain' which offers great transparency, visibility, and full knowledge of supply chain processes and data (PAT Research, 2021). The software suits the needs of small, medium, and large size organizations.

Features: The main attractions of this software are its power to optimize the supply chain, collaboration between businesses and effective order management. The software predicts, assesses, and controls the disruptions. Important strength of the software is to give a single view of inventory and order. Critical review of data and data interchange is easy with the help of the software. The software enables effective control of disruptions.

11.2.15 BluJay SCM

BluJay SCM has named its supply chain management software 'BluJay SCM'. The software is empowered with Cloud technology enabling hands-free customs and offers analytics, real-time, and velocity to the operations (PAT Research, 2021). The software suits the requirements of small, medium, and large size organizations.

Features: BluJay SCM replaces the traditional supply chain system by the Cloud technology-based supply chain management system, which reduces the risk, enhances efficiency, and takes advantage of the new opportunities. The focus of the software is on the areas: transportation management, warehouse management, commerce management, and network management. Shippers and logistics service providers take advantage of the software capabilities in the areas of custom regimes and global compliance. The commerce solution helps in quickly adding new suppliers and products for merchandising expansion and eliminating the out-of-stocks.

11.3 Ratings of the Top 15 SCM Software

PAT Research (2021), considering various factors, evaluated the top 15 SCM software in 2021 as discussed above. PAT Research (2021) worked out the ratings of the software based on its own evaluation as well as on the feedback received from the customers. The ratings based on the score by the Editor and the score given by the customers are shown in Table 11.1 (see next page).

11.4 Robotic Process Automation (RPA)

Robotic Process Automation (RPA) is a powerful software technology that connects the traditional supply chain practices with the digital age technology along with the next-generation technologies such as predictive sensors, machine learning, and artificial intelligence (AI). This technology integrates the human actions in performing the traditional repetitive tasks, and computer processes based on rules with the latest cognitive technologies such as machine learning algorithms or Artificial Intelligence (AI) (Deloitte, 2018).

This technology serves the need of transforming the traditional workforce into a digital workforce that can reduce human errors, make decisions based on judgment. Digital workforce can work with the humans of the traditional supply chains. The technology has been widely accepted by the industry.

11.4.1 Supply Chain Transformation

RPA technology helps carry out the traditional supply chain functions (plan, source, make, and deliver) and simultaneously also gives the real-time feedback. Digital workers join the digital and physical networks and contribute new connections between various activities of the supply chain (Figure 11.1). RPA has six elements: Synchronized planning, smart factory, intelligent supply, talent development, dynamic fulfillment, and connected customers in transforming the traditional supply chains into digital supply networks.

i. **Synchronized Planning**: RPA can automatically determine the reorder point and can place the order to ensure that the required material is received in the warehouse by the time reorder point is reached to avoid any stockout situation.

ii. **Connected Customers**: RPA along with other next generation technologies and digital procurement provides predictive service to consumer (S2C) strategic sourcing, peer-to-peer (P2P) automated transactional procurements, and efficient procure-to-pay process.

TABLE 11.1: RATINGS OF SCM SOFTWARE BY THE EDITOR & THE CUSTOMERS

Software Title	Supporting Technologies	Editor	Customers
E2open	Cloud based solutions	9.5	7.6
SAP SCM	Artificial Intelligence (AI) and Internet of Things (IoT)	8.8	7.2
Logility	Information not available	8.6	7.6
Perfect Commerce	Information not available	8.7	7.5
Oracle SCM	Cloud based technology	8.4	8.5
Infor SCM	Uses DOM and warehouse technologies	7.9	7.7
JDA SCM	Cloud based technology	7.9	8.5
Manhattan SCM	web-enabled portal facilitates real-time collaboration between the buyer, supplier, and trading partner.	7.8	5.2
Epicor SCM	Cloud based technology	7.8	7.5
Dassault Systems SCM	Information not available	7.8	7.8
Descartes SCM	Cloud-based, modular, and interoperable web solution	7.8	8
Highjump SCM	Cloud WMS	7.8	7.7
IFS	RFID technology	7.8	8
Watson Supply Chain	Information not available	7.7	6.5
BluJay SCM	Cloud powered portfolio of hands-free customs	7.5	8.1

iii. **Smart Factory**: RPA interacting with the internet of things (IoT) solutions offers faster response times and better understanding and consolidation of the data that IoT connected devices produce.

iv. **Intelligent Supply**: RPA managing the workflows with the addition of predictive analytics can transform equipment management from reactive to proactive to predictive.

v. **Talent Development**: Employees doing data-handling processes manually are made free with the use of process robotics. The employees can be deployed to carry out higher level processes, helping the organization to manage important and more complex works. In this way employees are trained and empowered with new skills.

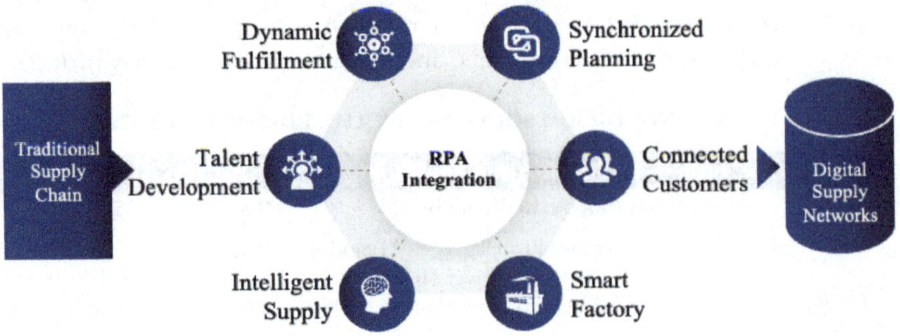

Source: Deloitte (2018)

Figure 11.1: RPA Integration of Traditional SCM with Digital SCM

The above activities with the use of RPA transform the traditional supply chain into digital supply networks. Organizations can implement new operational processes that best go in evolving technological capabilities and higher-demand environments in which they operate. The new efficiencies and operating models can effectively be adopted by governmental agencies involved in implementing large size programs to deliver more on time and on cost.

11.4.2 Advantages of RPA

The core benefits of RPA include:

i. Efficient fulfillment
ii. Improved customer value
iii. Higher network responsiveness and resilience
iv. Enhanced accuracy and better financial success
v. The transformation from the traditional processes to the digital era
vi. Employees can concentrate on strategic and more priority jobs. They are empowered with additional skills.
vii. Helps in transforming the traditional workforce into a digital workforce that can reduce human errors, make decisions based on judgment.

11.5 Key Trends in SCM Software

Cloud technology is used by majority of the software (Table 11.1). Other software, in addition to the Cloud technology, have other technologies also.

The software RAP developed by Deloitte (2018) is based on the integration of the next-generation technologies such as predictive sensors, machine learning, and artificial intelligence (AI). Consolidation, analytics, and automation are the key trends in the software used in supply chain management.

11.5.1 Consolidation

There are several supply chain management technologies that can contribute to enhancing the efficiency, productivity, and profitability of the company. The trend in the development of software is to integrate different technologies such as Cloud technology and artificial intelligence. The software, integrating as many technologies as possible, will be the market leader and the market choice.

11.5.2 Analytics

Supply chain analytics is one of the most potential areas in the coming days. It builds up the capabilities in different activity areas of the supply chain such as strategic planning, demand forecasting, procurement planning, inventory management, and warehousing management. Analytics is giving strategic information to the managers to make quick and strategically correct decisions helping in cost reduction, saving on time, enabling on schedule deliveries and the customer satisfaction. It empowers the company to measure performance of the supply chain based on Key Performance Indicators (KPIs). The power of analytics is so big that the customers will depend on the company as a most trusted business partner and would develop loyalty to the company.

11.5.3 Automation

Automation in the supply chain is especially important in all its activities whether it is demand forecasting, inventory management, procurement, store keeping, warehousing management, transportation, and handling of materials. Supply chain management has already covered good mileage in these directions but still more is yet to be covered. Automation is the visible trend in the days to come.

11.6 Summary

The Chapter introduces the concept of supply chain management software and the need for it in the supply chain management software. Objectives and functions of the software are highlighted. Further, the features of the supply chain management software and their benefits have been discussed. It is emphasized that supply chain management software brings efficiency, reduces the cost, increases competitiveness, and makes the organization profitable.

The chapter introduces top15 software of the year 2021 and discusses their features. In addition, it deals with an important supply chain software: RAP developed by Deloitte (2018) that shows that the trend is to adopt different technologies in the supply chain management software.

The Chapter discusses the analysis of ratings of the top 15 supply chain management software as determined by the editor as well as the ratings based on customer's review. Cloud technology is one of the most contributing technologies in all the software. Analysis has further brought out clearly that future software will be designed by including more technologies.

References

AnyLogic (nm). anyLogistix Approaches and Features. anylogistix-technical-datasheet.pdf. 1-11.

Deloitte (2018). The Path Forward: Operationalizing RPA to Automate the Digital Supply Network. Us-fed-operationalizing-rpa-to-automate-the-digital-supply-chain-network.pdf. 1-12.

PAT Research (2021). Top 15 Supply Chain Management Software. Paper posted on the website. www.predictiveanalyticstoday.com/top-supply-chain-management-software/

12

3D Printing

3D printing, also known as Additive manufacturing, is a manufacturing process developed since 1984 by Charles Hull (Deloitte, nd). 3D printing can create three-dimensional objects from digital data. 3D printing has already made inroads into the manufacturing sector right from product design to the production floor. Companies from different sectors such as engineering, automotive, aerospace, and healthcare are recognizing that 3D printing offers an opportunity to 'do things differently'. DHL (2016) identifies 3D printing as one of the major disruptive trends in the logistics industry soon. There is great potential for 3D printing to offer instant production and distribution models, enabling printing of complex objects with the help of a single printer. 3D printing is going to make a broad and deep impact especially in areas of spare parts manufacturing, customized parts, and end-of-runway 3D printing services.

Number of multinational companies have shown keen interest and have started to leverage this technology with attractive applications across many industries from aviation, healthcare, and food industry. Major initiatives were seen during 2016 with Mercedes-Benz Truck's announcement of its first 3D-printed spare parts service, HP launching its 3D printing initiative, and GE, BMW, and Nikon making a multimillion-dollar investment into the 3D printing start-up, Carbon1. These are only a few cases to name. McKinsey, a leading consulting firm, projects the growth of the 3D market ranging from $180 billion to $490 billion by 2025 (DHL, 2016). Wyman (2017) suggests 3D printing could represent a $400 billion manufacturing market by 2030. On a similar track, Gartner, the renowned information consulting company, considers that enterprise 3D printing is ready to make a dent and gaining widespread adoption (DHL, 2016). CIPS (2019) considers 3D technology as important in reducing the risk in investment in advance of the procurement by the firm. As an example, a complex or critical part of the plant such as a boiler in a power plant can be purchased at a minimum price by proper analysis in advance with the help of 3D modeling. The power company can save large costs involved in the procurement process.

12.1 3D Printing Technology

3D printing is a process used to prototype, produce small batches of customized products and individual components. A classic 3D printer operates in the same way as that of a conventional inkjet printer. Both types of printers receive printing information from a digital file. The inkjet printer applies ink to paper whereas 3D printers inject material in successive patterns to build a three-dimensional solid object.

12.1.1 Working of 3D Printing

The working principle of a 3D printer is explained in Figure 12 1:

i. A laser beam is sent from a laser source to solidify the material
ii. The elevator lifts and lowers the platform to help lay the layers
iii. The 3D object is made with the material contained in the vat
iv. The 3D object is built as parts are layered on top of each other
v. One or more materials which may include plastic, resin, titanium, polymers, gold, and silver are used in advanced 3D printers.

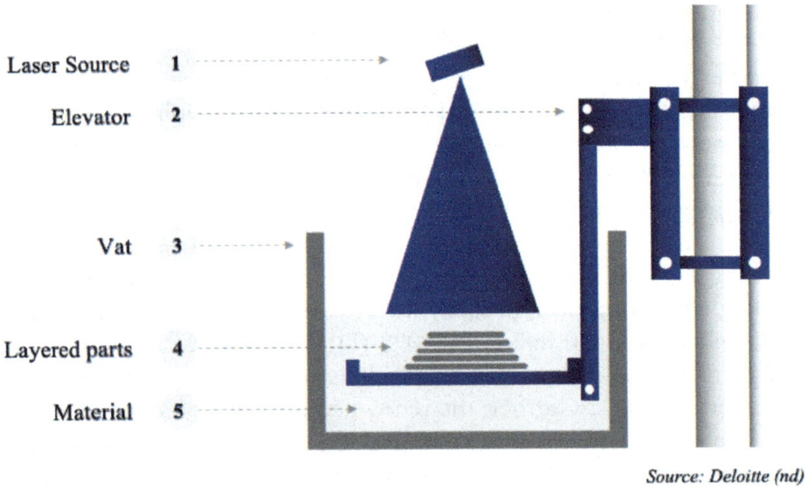

Figure 12.1: Working Principle of 3D Print

Injection of material layer by layer differentiate the 3D printing manufacturing from the traditional manufacturing process where the additional material to the part is added by molding or cutting or bending materials (DHL, 2016). A 3D printer is the hardware to produce solid objects out of the digital model and feed material.

12.1.2 Traditional Manufacturing Vs. 3D Printing

There is noticeable difference between traditional manufacturing and 3D printing. In case of traditional manufacturing, materials are purchased from different suppliers located at different places and shipped to centralized factories where they are handled/processed and assembled into the finished product. Once an order is placed, the finished goods pass-through number of stages in the supply chains such as warehouses to stores or directly to the end-customers.

The case of 3D printing is significantly different from the traditional manufacturing process. 3D printing reduces the complexities of traditional manufacturing to a great extent. A single 3D printer can produce a variety of products. 3D printing eliminates many processes and handling in the supply chains in comparison to traditional manufacturing. Companies, with the help of 3D printing technology, adopt on-demand de-centralized production of the finished goods. This change in the manufacturing methodology led to great savings from production and logistics costs. Further savings are achieved from the reduction of waste in the production and adoption of sustainable business practices.

Airbus has an impressive example in redesigning its hinge brackets that attach parts to an aircraft's fuselage using 3D printing (DHL, 2016). Airbus could reduce the material waste by 25% through 3D printing in comparison to traditional casting methods (Figure 12.2).

Note the following key items about the image shown in Figure12.2 above:

i. Left bracket made by traditional manufacturing.

ii. Right bracket made by 3D printing.

iii. Reduction in weight by 3D printing: 10Kg.

iv. The weight of the airplane was reduced by 10 Kg with the new design of the bracket with the use of 3D printing. In addition to these impressive savings, the Airbus could achieve a reduction of 40% in CO2 emissions over the lifecycle of the bracket.

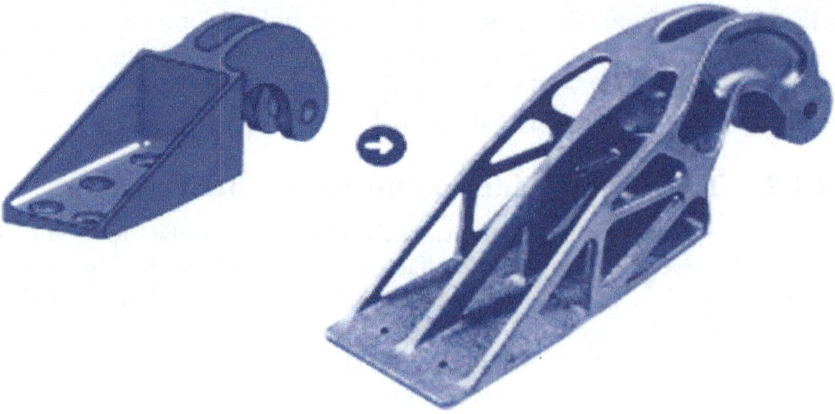

Source: Gartner (2016)

Figure 12.2: Traditional Vs. 3D Print Design of Bracket

12.1.3 3D Printing - An Attractive Option?

Key factors that make 3D printing an attractive option include the following (DHL, 2016):

i. 3D printing involves a smaller number of steps in designing, prototyping, and manufacturing highly complex and/or customized products.

ii. On-demand and decentralized production strategies of 3D printing enable delivery in a shorter time.

iii. Logistics and production costs are lower in case of 3D printing due to reduced shipping and storage, elimination of import/export from the localized production, and no need for new production tools and molds.

iv. Lower amount of material and energy in production lead to higher sustainability and efficiency in production.

12.1.4 Impact of 3D printing on Global SCM

3D printing is going to disrupt every aspect of global supply chain management such as warehousing, inventory control, and locating the manufacturing facilities near or at the center of consumption.

i. **Location of the Manufacturing Units:** Multinational companies prefer to locate production units in a specific country to optimize cost by economizing on the input cost of the production process, which may include cost of raw materials, cost of labor, tax incentives offered by the government, lower cost of energy and water, land made available at subsidized rate, and many more incentives. Malaysia, China, India, and Indonesia are the countries with lower cost of labor, an important factor in the cost of production. Refratechnik, a German firm and leader in the manufacturing of the refractory materials used in cement plants, had set up its manufacturing unit in China due to cheap labor. Availability of raw materials is another important factor in selecting the site for manufacturing the product. Limestone, bauxite, and gypsum are important raw materials to produce cement. Cement plants are set up near raw materials to save the cost of transportation of the raw materials. Since the location of cement plants is usually in remote areas, use of 3D printing will be a very useful tool in cost reduction of spare parts which can be made easily at site.

Dell, Samsung, Nokia, Onida, Siemens, Magotteaux have set up their manufacturing facilities in India due to closeness to the market, and low cost of labor. Most of these companies have set up their production facilities in locations where they could save on costs of labor, materials, and shipping.

ii. **Reduction in Shipping Costs**: Shipping is often the most expensive element of operations. Shipping cost is increased across the supply chains, as raw materials are to be shipped to the manufacturing units, the products manufactured are to be shipped to the assembly plants, and the finished products are to be shipped to the market. 3D printing plays an important role across the supply chains due to production at lower cost due to proximity of the manufacturing facility near the consumers. This has a great impact on the global shipping industry.

iii. **Dynamic Optimization of Inventory of Spare Parts**: Planes, trains, trucks, and cars undergoes continuous wear and tear and need spare parts specific to each make and model for replacement. Wide range of products have hundreds of millions of spare parts that are kept in storage across the world. Products may range from production plants, cars, watches, x-ray machines, and washing machines. Inventories in the warehouses at any time have many items which will rarely be used, and some may never be needed. According to DHL (2016), the share of excess inventories can sometimes exceed 20%. The inventory carrying cost of the companies becomes very high resulting higher cost of production. With the higher cost of production, the products are no longer competitive in the market. It is costly for companies to store the slow moving or non-moving stock leading to build inefficiency into the supply chain, as excess parts may never come into use.

Spare parts sector is one of the first areas to be disrupted with the use of 3D printing. Companies need not store spare parts physically in the warehouse any more as the spare parts can be printed easily on demand. The spare parts can be delivered to the customers rapidly. Logistics providers can create an integrated network of 3D printers to instantly print and deliver spare parts on demand. Companies can have a virtual warehouse with virtual print files of spare parts securely stored in the software databases acting as virtual warehouses.

Users can simply search for the required spare parts and send the file to the nearest 3D printer. Mercedes-Benz Trucks has already adopted this strategy (DHL, 2016).

iv. **Manufacturing of Individual Direct Parts**: 3D printing offers a source for customized parts and gets them quickly to the point of use. Companies can form a network of 3D printers with logistics providers which can work as a small micro factory. These printers can be in regional warehouses or local distribution centers and can be linked with an OEM's label. This concept can be very useful for the healthcare sector which is already using 3D printing in several applications achieving quality standards comparable with traditional manufacturing methods. All types of healthcare facilities are collaborating with medical companies and logistics providers to build end-to-end 3D printing services that reduce cost and are simple. The logistics provider can provide fast, safe, and secure delivery of each part to the right place and at the right time.

A patent for a truck fitted with 3D printers has been filed by Amazon for manufacturing products on the way to a customer destination (DHL, 2016). This could enable companies to make parts very close to demand and thereby drastically reduce the lead time to customers

v. **Product Postponement Services**: Customers can select aspects of the design, material, shape and size, packaging, and product functionalities in customizing the item and availing the postponement service by using 3D printing. Local distribution centers can keep almost–finished goods as well as 3D printers that carry out a variety of customization functions before delivering the product to the customer.

vi. **End-of-Runway Services**: Logistics providers are often under pressure to act fast with minimum response times and speed to market for time-sensitive shipment of critical parts. Sector-specific service offerings and integrated return and repair services leverage the use of end-of-runway services, an integrated logistics solution with the location of specific warehouses close to the important airport hubs. The high speed of 3D printing enables end-of-runway hubs to produce faster the parts that need to be put into operation in the shortest possible time frame. Essential parts can be printed and delivered immediately for replacement in short lead times enhancing the levels of customer satisfaction.

Spare parts for renewable energy plants, brake pads for wind turbines and fuel nozzles for aircraft are some of the industrial applications where 3D printing can make significant contributions. Consumer sector can also use 3D printing in end-of-runway logistics to improve repair operations that are covered under warranty.

Fast Radius of UPS has strategically set up its 3D printing factory very close to the UPS global air hub at a distance that can be covered in a few minutes only. The close end-of-runway location enables the production of parts up to the 1 a.m. pickup time that can be delivered anywhere in the U.S. the next morning.

12.1.5 Advantages of 3D Printing in SCM

3D printing has emerged as one of the most promising technologies in supply chain management. Some of the advantages of 3D printing are:

i. 3D printing would require less raw materials for production and would result in less waste due to "additive manufacturing".

ii. Single item can be produced economically without spending on the mold and tooling costs as required in traditional manufacturing.

iii. 3D printing encourages increased innovation flexibility motivating firms to experiment with new prototypes/products of complex shapes and structures very quickly. Such innovations do not need large investments in expensive machinery, tools, and components usually required in traditional manufacturing.

iv. The overhead cost in investing in inventory and the warehouse is reduced as the items can be printed when needed.

v. 3D printing offers manufacturing of parts in small mobile facilities near the place of consumption leading to reduction in transportation cost, lead time.

vi. 3D printing helps in the use of local sourcing in place of global sourcing leading to reduced lead time, and transportation cost.

vii. 3D printing simplifies inventory management and makes it economical. It enables production of the simple product in small lots at different distributed locations resulting in lower shipping cost.

viii. 3D printing can produce customized products in small lots on demand in contrast to large scale production and reduce the need for storing the finished products.

ix. Automotive manufacturing companies such as Ford and General Motor used 3D printing in producing prototypes of automotive spare parts at reduced cost.

x. 3D printing helps to reduce the tooling cost.

xi. 3D printing supports pull deliveries from the suppliers in place of push deliveries. The inventory carrying cost is reduced due to Just-in-Time (JIT) inventory.

xii. The rejection risks are reduced, helping reverse logistics operations. "Returns" from customers are reduced because of better customer service (i.e., personalized products) and easier management of reverse logistics because of closeness to end consumers

xiii. 3D printing helps in spare parts management from low diversification of product design and possibilities of their production at site.

xiv. Damage to the environment is less with 3D printing.

12.1.6 Large Scale Deployment of 3D Printing

Consumer demand is relatively in smaller volumes for the highly personalized complex products. In view of this, industry is likely to face certain barriers before realizing the projected benefits at a large scale. Some of the potential barriers in a large-scale deployment of 3D printing are:

i. 3D printing needs higher energy, 50 to 100 times more than injection molding on products of the same weight (Wyman, 2017). 3D printing is, therefore, costly for large-scale manufacturing. It poses potential health risks due to melting plastics.

ii. It is simpler to copy an item produced on a 3D printer. This raises the issue of infringement of intellectual property rights. According to a Gartner report, 3D printing will cost $100 billion per year in IP infringement.

iii. With the currently available technology, 3D printing offers limited opportunity in producing items of all sizes.

iv. Traditional manufacturing enjoys the economies of scale and takes shorter time as compared to 3D printers taking time to print products.

v. 3D printing, despite some of the above-mentioned barriers, is becoming a popular digital technology, and has a very significant impact on supply chains.

12.2 Summary

3D printing has emerged as one of the fastest growing technologies that has been adopted in many sectors like engineering, automotive, aerospace, and healthcare. The Chapter explains the concept of 3D printing followed by its working principle. This is followed by the difference between traditional manufacturing and 3D printing. The Chapter discusses the key factors that make the 3D printing an attractive technology, impact of the 3D printing on the global supply chains, the advantages of 3D printing and barriers in large scale deployment.

References

CIPS (2019). Digitalization in Procurement and Supply. Chartered Institute of Procurement & Supply. Melbourney.1-20.

Deloitte (nd). Disruptive manufacturing: The effects of 3D printing. Deloitte LLP, Canada. 1-18.

DHL (2016). 3D Printing and the future of Supply Chains: A DHL perspective on the state of 3D printing and implications for logistics. DHL Customer Solutions & Innovations, Germany. 1-31.

Gartner (2016). What 3D Printing Means for your Supply Chain. Gartner Inc. http://www.gartner.com.

Wyman, Oliver. (2017). 3D Printing Is Already Starting to Threaten the Traditional Spare Parts Supply Chain (forbes.com) http://www.logisticsmgmt.com/article/ups_rolls_out_plan_for_full_scale_on_demand_3d_printing_manufacturing_network.

13

Digital Transformation in Supply Chain Management

Digital transformation is the most talked about subject in todays' business environment. Every organization, whether in business, industry, Government, or education is keen to go for 'Digital Transformation' to improve accessibility to fast information, transparency in various operations, and improve efficiency. European Union (EU), supporting the digital transformation of business and society, initiated Digital Transformation Monitor and Digital Transformation Scoreboard with the aim to measure progress on digital transformation in EU countries (European Commission, 2019). On similar lines, the Organization of Economic Cooperation and Development (OECD) initiated a global project in 2017 "Going Digital: Making the Transformation Work for Growth and Well-being" to create better understanding about digital transformation taking place around the world and help various countries to be digital and data-driven (OECD, 2018).

Digital transformation of the supply chains has become a need rather than a choice. This has been clearly realized during one of the most difficult situations of COVID-19 period where everything got disrupted. Top management has realized that adopting the latest technology can only sail them through the storm. Digital leaders were able to absorb the initial shocks of the pandemic COVID-19 by deploying technology and managing the operations by remote workforce engaging the customers via contactless deliveries. Many of the digital leaders took advantage of the situation and searched for new opportunities for the organization through technology supporting new business models and new innovative ways of working. However, successful digital transformation needs a cultural shift in the way of thinking, working, and approach in solving the problem. Organizations invested in digital solutions courageously to face the disruption created by COVID-19.

13.1 What is Digital Transformation?

Digital Transformation (DT) has attracted the attention of many authors who have tried to define DT as perceived by them. Verina and Titko (2019) compiled the definitions of digital transformation as provided by various authors. Table 13.1 presents the definitions suggested by Schwertner (2017), Ismail, et al. (2017), Deloitte (2018), Bloomberg (2018), OECD (2018), and European Commission (2019). In addition to the compilation of definitions suggested by other authors, Verina and Titko (2019) developed the concept of Digital Transformation. The definition of DT based on the analysis of Verina, and Titko (2019) is presented in Table 13.1.

Based on these definitions, the following definition of Digital Transformation is recommended:

> *"Digital transformation is the integration of organizational changes, with multiple digital technologies along with development of skills through training and improved recruitment process of the people to reach superior performance, competitiveness of the organization, and meeting the aspirations of the customers. The applications of the digital technologies help in designing new business models, developing new systems, processes, and human resources with new skills to match the new technologies. DT brings in a cultural change in the organization with new decision-making processes. The resultant outcomes of digital transformation include more profitable revenues, greater competitive advantages, and higher efficiency".*

13.2 Benefits from Digital Transformation

Organizations can achieve several benefits from the digital transformation (TEKsystems, 2021), which include:

i. Integration with other business systems and ease of operations with other technologies and software in use

ii. Analytics and business intelligence reporting

iii. Security

iv. Predictive analytics

v. Virtual workspaces / remote work capabilities

vi. Data privacy

vii. Scalability

viii. Edge-to-cloud connectivity

TABLE 13.1: DIGITAL TRANSFORMATION DEFINED BY OTHER AUTHORS

Source	Definition
Schwertner (2017)	"The application of technology to build new business models, processes, software and systems that result in more profitable revenue, greater competitive advantage, and higher efficiency".
Ismail, Khater, and Zaki (2017)	[Digital transformation is a] "process through which companies converge multiple new digital technologies, enhanced with ubiquitous connectivity, with the intention of reaching superior performance and sustained competitive advantage, by transforming multiple business dimensions, including the *business model*, the *customer experience* (comprising digitally enabled products and services) and *operations* (comprising processes and decision-making), and simultaneously impacting *people* (including skills talent and culture) and *networks* (including the entire value system)."
Deloitte (2018)	"Digital transformation is the use of technology to radically improve the performance or reach of an organization. In a digitally transformed business, digital technologies enable improved processes, engaged talent, and new business models."
Bloomberg (2018)	"Digital transformation requires the organization to deal better with change overall, essentially making change a core competency as the enterprise becomes customer-driven end-to-end. Such agility will facilitate ongoing digitalization initiatives but should not be confused with them."
European Commission (2019)	"Digital transformation is characterized by a fusion of advanced technologies and the integration of physical and digital systems, the predominance of innovative business models and new processes, and the creation of smart products and services."
Verina & Titko (2019)	Digital Transformation includes "organizational change", "cultural transformation", and "customer-centric approach". People are an essential element and more important than any other thing. DT takes care of internal communication with staff members, training, motivation of the staff as well as rebuilding the recruitment program. Needs of the organization in digital skills be identified and built up to match the organizational strategy.

13.3 Digital Transformation in SCM

Industry 4.0 has been the driving force in digital transformation of business all over the world, but the pace of digital transformation has been slow due to many hurdles in the process. COVID 19 disrupted every aspect of life forcing government, industry, businesses, and educational institutions to change their working pattern and business strategies. Imposed lockdowns in various countries pushed remote working leading to almost total remote workforce. Companies used the forced situation to use digital technologies like Cloud, data analytics, AI, and automation to the maximum extent to sustain the business.

Market pressures resulting from challenging situations such as pandemic act as the catalyst for digital transformation efforts. Customer is the focus in any strategic move related to digital transformation. Enhancing the customer experience and engagement are the top considerations in digital transformation. Speed-to-market and innovation are other important aspects for the organizations. Contactless delivery in a limited face-to-face interaction in the current business environment is of great importance.

TEKsystems (2020), just before the pandemic forcing lockdowns, conducted an online survey about the state of digital transformation during the period November to December 2019 covering 510 participants at 'C' levels from the industry. In view of COVID-19 showing its impact in the beginning of 2020, TEKsystems (TEKsystems, 2021) again surveyed the impact of pandemic on digital transformation efforts during November to December 2020 with more than 900 participants. TEKsystem's surveys showed that:

i. Majority of the companies have aligned their work to support remote working. Eighty percent of the organizations expect 50% or more of their workforce to work remotely at least during the remaining part of 2020.

ii. Organizations are planning increasing use of the following digital tools during the new normal:

- Instant messaging tools (94% participants)
- Video conferencing (74% participants)
- File sharing tools (56% participants)
- Project management tools (21% participants)
- Payment and invoice tools (15% participants)
- Sixty percent organizations inform that the pandemic increased the need for urgency in initiating digital transformation. COVID-19 has been responsible for increasing the plans for spending on technology in 2020.

The goals for digital transformation may include the following:

i. Enhance customer experience and engagement

ii. Increase operational inefficiency

iii. Transform the IT system by replacing or upgrading the systems

iv. Products or services to reach the market with great speed in the shortest time.

v. Encourage innovations and help in their implementation quickly

vi. Develop new business models and revenue streams

vii. Develop new products and services

viii. Strengthen cybersecurity

ix. Retain talent

COVID-19 pandemic changed the belief that technology deployments must be slow and bulky. Implementation of the new business model should be quick for value addition, modifying, and moving forward. Establishing a new way of working is a critical component for digital transformation success. It needs training and development. Pandemic has created new demand on the organizations to manage remote workforce.

Outbreak of pandemic lead to increasing use of different technologies, which include videoconferencing, remote learning /training, Cloud computing (i.e., IaaS, PaaS, SaaS), Virtual workspaces, collaborative applications, data analytics, secure remote access, employee engagement, process automation, data/information, enterprise social network, and mobile devices. Organizations are balancing between the digital initiatives to sustain them in near term against the holistic modernization efforts. Digital leaders are taking the opportunity to act and transform their business.

Hackett Group (2020), in its survey 'Key Issue - 2020', identified four areas for critical development related to objectives of the enterprises and two areas of critical developments related to the objectives of supply chains (see Figure 13.1 above).

Enterprise digital transformation and enterprise analytics capabilities are critical to achieve the objectives of the enterprise. In the case of supply chains, the survey found that improvement of business processes and improvements of analytical capability of supply chains are the critical areas of attention. Organizations should focus on implementation of digital technologies to achieve the objectives of the enterprises as well as that of the supply chains.

Source: Hackett Group (2020)

Figure 13.1: Achieving Objectives of Enterprises and Supply Chains

Analytics capabilities are the needs, both in terms of broad and in-depth, to match with the objectives of both the enterprises and supply chains.

13.4 Challenges in Driving SCM Transformation

Transformation of traditional supply chain management into digital supply chain management faces several problems. Despite the determined efforts in implementation of digital supply chain management, common problems faced by the industry include the following:

i. **Lack of Understanding of Digital SCM**: Awareness about Industry 4.0 and concept of digital supply chain management are not understood by many professionals in the industry and business. This is one of the important drawbacks in carrying forward digitalization of the supply chain management. There is a need to create this awareness and the benefits of transformation of traditional approach to digital supply chain.

ii. **Resistance to Change**: Many organizations have great resistance to change. Fear of lack of skills in the existing staff deter them to take a step forward in adopting new technology.

iii. **Inadequate Automation**: Many small companies do not have computers and the knowledge of using computers. At the same time many of the customers do not either have computers or mobiles or do not know the use of appliances such as computers, mobile phones and iPads. Digitalization of supply chain management can support these facilities on a large scale especially in small cities and villages.

iv. **Poor Internet Connectivity:** Internet services are not of satisfactory level. Non-availability or interruptions make the digitalization difficult. Confidence level is still lacking.

v. **Over Commitment:** Over commitment to take up many digitalization projects at the same time is difficult for an organization. There may be too many technology priorities which may be competing. Supply chain managers fail to prioritize the projects from a long list. The net result is that digitalization starts getting delayed.

vi. **Lack of Funds:** Digital technologies need large amounts of funds. Organizations do not have enough funds to initiate digitalization programs. Organizations should make practical strategies to create funds for the new technologies.

vii. **Data Issues:** It is difficult to get correct and reliable data.

viii. **Security and Compliance:** Security concerns are increasing day by day. Hackers are active in breaching the data protection tools.

ix. **Limited Resource Bandwidth:** Qualified staff is in shortage. Digitalization efforts will not be effective if skilled staff is in short supply.

x. **Talent Management:** There is a shortage of trained staff to install new technologies. Besides the availability of the skilled staff, there is a need for trained manpower. Execution of the digitalization program will not be smooth in absence of trained staff in specific technology.

xi. **Executive Sponsorship:** Digitalization program cannot be implemented without the support of senior management. Involvement of Chief Executive is very essential for successful execution of the digitalization program.

xii. **Organizational Alignment:** Organizations face different complexities such as inappropriate organization structure, overlapping of responsibilities between managers, unclear hierarchy or reporting system, and lack of policies. Organizations may be having employees with siloed mindset who do not interact with their peers and are not able to solve the issues or have adamant behavior not caring of the views of others. There are staff members who are arrogant and at times behave improperly with other members. Senior management must keep an eye on such problems to keep them under control.

xiii. **Organizational Change Management:** Every organization should have flexibility in its business processes. Operating model of the organization should be flexible to meet every situation. COVID-

19 has shown this very clearly. Organizations with flexibility in their operating model could make quick changes and align themselves with the changing situations. Rigid organizations could not operate in an uncertain pandemic environment due to uncertain resource capacities and resource availability.

13.5 Agenda for Transformation in SCM

Hackett Group (2020), based on the findings of the 'Key Issue' survey 2020, developed an agenda for the transformation of the supply chain. The agenda items, according to the priorities, include the following:

i. Analytics and Information Management.

ii. Human Capital Management

iii. Service Design.

iv. Technology

v. Organization and Governance

vi. Service Partnering

13.5.1 Analytics and Information Management

This is the most important aspect in the transformation of supply chain management (Hackett Group, 2020). Supply chains have adopted advanced analytics in many ways. More opportunities are evolving across the supply chains for adoption of advanced analytics.

13.5.2 Human Capital Management

Human capital supports every aspect of the operations. Companies are focusing on delivering higher output with a smaller workforce. Competent workforce can be built by taking following actions:

i. Attention be given to retain, engage and develop the existing employees and at the same time recruit new employees.

ii. Identify the gap between the available skills and the needed skills to develop digital products and services offered.

13.5.3 Service Design

Service design focuses on the processes to meet the customer's needs and organizational goals. Improving distribution/logistics costs, supply and operations capabilities, and reduced risks are of high priority.

13.5.4 Technology

With the development of digital technologies, organizations are focusing on process improvements, increased efficiency, and reduced costs. Supply chain analytics will play an even more important role in the future in transportation planning, MRP, supply chain visibility, and risk management. Various technologies like Cloud technology, Artificial Intelligence (AI), simulation, robotics will lead to fast digital transformation of supply chain management.

13.5.5 Organization and governance

The strain on the overall supply chain management has suddenly increased with the unforeseen high demands from COVID-19. The responsibilities with increased workload are going to be part of the supply chain management. The maximum impact will be on corporate and in individual business units.

13.5.6 Service Partnering

Service partnering or collaboration with partners in pursuing the organizational goals is high on the agenda for transformation of the supply chain and its digitalization. There will be more outsourcing and partnerships to meet the demand with less internal workforce. These moves are to reduce cost and at the same time offer better performance and customer satisfaction.

13.6 Strategy for Digital SCM Transformation

Digital transformation is an incredibly challenging task. It calls for good planning and execution. Following steps will help in digital transformation:

i. Develop vision for digital transformation focusing on customer's needs.

ii. Identify the digital weak areas in the company and prioritize them.

iii. Identify digital technologies to be introduced.

iv. Develop a comprehensive plan to cater to the long-term needs of the organization. The comprehensive plan should include technology planning, material planning, manpower planning, facility planning, information technology planning, and any other needs for implementation of the plan.

v. Prioritize the implementation of selected technologies

vi. Select the place to house the facilities of digital transformation.

vii. Obtain consensus from all the departmental heads about the digital transformation plan.

viii. Listen to the points of view of everyone and remove any apprehensions in anybody's mind.

ix. Communicate across the organization about the project, its aims, details, execution plans, and expectations in the execution of the project from different sections and key actors.

x. Involve the whole organization in the execution and success of the project.

xi. Install reliable internet facilities or strengthen the existing internet facility which should be fast and dependable.

xii. Involve the board members to facilitate approval of the board. Chief executive should actively participate right from the beginning till completion of the project.

xiii. Identify the focal officer to execute the project.

xiv. Make a team of staff responsible for digital transformation.

xv. Recruit the staff to fill the gap in the skills and competency areas.

xvi. Conduct awareness programs, dedicated training programs for officials responsible for the execution of the project and the officials who will be using the technology.

xvii. Train the concerned staff across various departments with the operations, maintenance of the system, and troubleshooting.

xviii. Cultivate a culture of acceptance of change and agility.

xix. Sufficient funds are critical to implement the project. Start the project only on assurance of funds.

xx. Provide expertise knowledge from external sources if internal expertise is not available or is not free for the project. Establish a contract with a competent consultant/consulting firm having experience in similar projects.

xxi. Identify competent supplier/agency to supply, install, and commission the system.

xxii. Inform all the departments about the project to ensure cooperation from every department regarding sharing the information as required.

xxiii. Monitor the project with the help of the latest tools such as PERT and CPM.

xxiv. Evaluate the installed digital technologies. Remove any deficiency or problem noticed.

xxv. Establish a risk management system to protect data and information. The digital risk management system should be capable of protecting the system from cyber-attacks.

xxvi. Align the vision, culture, and structure of the organization. Technologies like AI, cloud, edge computing or blockchain are behind digital transformation but they are just enablers to move the organization forward.

xxvii. The strategy to adopt these technologies is critical, particularly in view of two important points: people and culture. Changing the mindsets and introducing new ways of working are the main aspects that ultimately define successful and sustainable digital transformation.

13.7 Success in Digital SCM Transformation

Digital success relates less to technology and more to people. Technology is mainly used to enable innovation, automate processes, and offer insights that lead to growth. It is the employees that must accept the change and adopt a total change of mindset and ways of working that use technologies like cloud, data analytics and AI. Do not ignore the importance of changes in culture and behavior. Digital leaders successfully integrate new ways of working and collaborate with appropriate training and development programs to nurture new behaviors and interlink them with the tools and technologies.

13.8 Accelerating Digital SCM Transformation

Technology adoption can create clear competitive advantages. But adoption is only part of the equation. Digital leaders accelerate their digital transformation journey with a clear vision and framework, where technology plays the role of enabler. Digital leaders identify the business problems they wish to solve and determine the desired outcome. They clearly define success metrics, establish a roadmap with key milestones, and finally focus on continuous improvement, adjustments, and capturing feedback to deliver results.

13.9 Summary

The Chapter discusses benefits from digital transformation and the definitions of digital transformation as presented by different authors. The author then suggests a smart definition of the digital transformation. He then discusses the scenario of digital transformation of supply chain management and the findings of the surveys by TEKsystems before and during the

COVID-19. The findings suggest that an organization can move in diverse ways for digital transformation such as focusing on optimization to improve efficiency or focusing on disruption and boosting the business agility. There is no fixed rule to adopt a particular approach except that the interests of the customers are always the priority. The goals of the business including the value the goals can serve the needs of the customers are the most important things. The Chapter brings out the critical development areas for digital transformation and the problems in digital transformation of supply chain management. The agenda for the transformation of supply chain management, strategy for digital transformation of SCM, elements of success in digital transformation, and how to expedite the digital transformation are discussed in detail.

References

Bloomberg, J. (2018). Digitization, digitalization, and digital transformation: confuse them at your peril. Retrieved from https://www.forbes.com/sites/jasonbloomberg/2018/04/29/digitization-digitalization-and-digital-transformation-confusethem-at-your-peril/#2dd1ce842f2c

Deloitte. (2018). Digital enablement turning your transformation into a successful journey. Retrieved from https://www2.deloitte.com/content/dam/Deloitte/ie/Documents/Technology/IE_C_HC_campaign.pdf

European Commission. (2019). Digital transformation. Retrieved from https://ec.europa.eu/growth/industry/policy/digitaltransformation_en

Hackett Group. (2020). The Digital Supply Chain's Evolving Role in the Enterprise Agenda. *Procurement Executive Insight - The Hackett Group.* 1-8.

Ismail, M. H., Khater, M., & Zaki, M. (2017). Digital business transformation and strategy: What do we know so far? Retrieved from https://cambridgeservicealliance.eng.cam.ac.uk/resources/Downloads/Monthly%20Papers/2017NovPaper_Mariam.pdf

OECD. (2018). Going digital in a multilateral world. Retrieved from https://www.oecd.org/going-digital/C-MIN-2018-6-EN.pdf

Schwertner, K. (2017). Digital transformation of business. *Trakia Journal of Sciences.* 15(1). 388-393. https://doi.org/10.15547/tjs.2017.s.01.065

TEKsystems. (2020). 2020 State of Digital Transformation: Building a Framework for Digital Success. *TEKsystems Global Services, LLC.* 1 - 40.

TEKsystems. (2021). 2021 State of Digital Transformation: Building a Framework for Digital Success. *TEKsystems Global Services, LLC.* 1 - 62.

Verina, Natalja, and Titko, Jolena (2019). Digital Transformation: Conceptual Framework. *International Scientific Conference* 'Contemporary issues in Business Management and Economics Engineering, 2019', 9-10 May 2019. Vilnus, Lithuania. 1-10.

Section III

Supply Chain Risk Management Post COVID-19

14

Risk Management in Digital Supply Chains

Supply Chain Management connects the manufacturer, vendors, many third-party companies, government and regulatory agencies, and the customers. Number of enterprises are interconnected between the starting point and the end point in the supply chain. Industry 4.0 has introduced the concept of Digital Supply Chain Management (DSCM) integrating the objective number of technologies such as Cloud technology, Internet of Things (IoT), Artificial Intelligence (AI), Big Data, Robots, RFID, and others. The objective is to introduce transparency, speed in operations making the supply chain more efficient and productive. Collaboration is one of the most important strategies to make the supply chain successful with real time information sharing. The extensive interaction with real time information sharing has been possible with extensive use of sensors. The introduction of the digital technologies, on one hand, has made decision making extremely fast and accurate and on the other hand has added risks due to sharing of information knowingly and unknowingly. This has led to new risks called as 'Cyber Supply Chain Risk' in the supply chains. Different authors have differently defined cyber chain risks as presented in the following section.

14.1 Cyber Supply Chain Risks

Different authors have differently looked at cyber supply chain risks. The cyber security risks have a broader scope than the conventional risks. The scope of risk is broadening over the time as the online activities of the companies are enlarging. The different definitions of cyber risks as given by different authors are presented in Table 14.1.

TABLE 14.1: CYBER SUPPLY CHAIN RISKS AS DEFINED BY DIFFERENT AUTHORS

No.	Definition	Author
1	SC risk is "the variation in the distribution of possible SC outcomes, their likelihood and subjective values."	March and Shapira (1987)
2	The possibility and effect of mismatch between supply and demand. Further, SC risk is any risk to material, product, and information flow from the original supplier to the delivery of the final product.	Juttner et al. (2003)
3	The probability of loss arising because of incorrect, incomplete, or illegal access to information.	Faisal et al. (2007, p. 679)
4	Degradation or disruption to a supply chain's infrastructure or structural resources by exploitation of IT vulnerabilities by threats within an organization, within the supply chain network, or in the external environment is known as cyber supply chain risk.	Smith et al. (2007) Electronic copy available at: https://ssrn.com/abstract=3426030 Supply Chain Risk
5	"IT security incidents occur when a threat directed against an organizational asset causes a compromise in one (or more) of three areas: confidentiality, integrity or availability (CIA)."	Deane et al. (2009, p. 5)
6	The potential deviations from initial overall objective that, consequently, trigger the decrease of value-added activities at different levels.	Kumar et al. (2010)
7	Any crime facilitated or committed using a computer, network, or hardware device; in particular, the computer or the device may be the agent, facilitator, or target of the crime is considered a cyber security risk which takes place in virtual or non-virtual places. A cyber-security risk leads to loss of functionality, connectivity, performance, or capacity.	Cebula and Young (2010), Urciuoli et al. (2013, p. 51)
8	Risks from the internet which are maliciously introduced with unwanted functions, or that subvert the design, product, or integrity of the system are the Cyber Supply Chain Risks. Source: The Ike Skelton National Defense Authorization Act for Fiscal Year 2011, (Section 806).	Wallace (2016)

Cyber security risk is different from conventional supply chain risks. Cyber risk is anonymous as it can remain undetectable until it impacts the business.

Considering the various aspects of the definitions given by various authors at different times, the Cyber Supply Chain Risk in the current context is defined as below:

> *"Cyber Supply Risk is the result of the activities from the internet and are introduced maliciously through unwanted functions that negatively impact the design, product, or the integrity of the system. A computer, network, or hardware device such as sensors used in digital technologies are the agents committing the crime in virtual or non-virtual place or both. The malicious act*

results in loss of functionality, connectivity, or capacity of the organizational assets and impairs the confidentiality, integrity, or the availability. The IT infrastructure of the organization is affected by the malicious act knowingly or unknowingly within the organization, within the supply chain network, or the external environment known as Internet of Things (IoT). The malicious act is carried out by incorrect, incomplete, or illegal access to the information system by an internal employee, and external source, or both."

14.2. Sources of Cyber Supply Chain Crimes

Sources of cyber supply chain crimes can be linked to some of the latest technologies used in digital supply chains. According to KMPG (2018), some of 'the most-in-demand supply chain technologies enabling improvements include:

i. Robotics

ii. Optimizing warehouse flows and processes

iii. Auto replenishment to home

iv. Facial recognition technology

v. Driverless vehicles, drones, and on-board technology

vi. Smart labels, QR codes and

vii. Blockchain technology

The latest technologies while improving the performance of supply chains, they also make the companies very vulnerable. The interconnected devices in the supply chains allow access to external parties such as customers in the organization. The organizations are thus open to many types of 'attack vectors' into their systems.

All the emerging supply chain technologies use access points, sensors, and scanners. KMPG (2018) estimated that 30 billion connected devices as part of the global IoT base, up from 14.5 billion in 2015. These devices are often physically dispersed and are not always integrated with the organization's IT security system design. This exposes organizations to significant supply chain security risks. KMPG (2018) estimated that up to 80 percent of security breaches occur through the supply chains. The organizations need to be protected from these security risks.

The new emerging digital supply chain technologies bring sophistication with them. When interconnected in the networks, these smart technologies add sophistication and complexity and introduce potential risks and impact of cyber-attacks. The need is to understand and manage the interconnected supply chains and their points of vulnerability and weaknesses to prevent and manage issues. Cyber risks are to be considered with a holistic perspective. Cyber security has become a business issue drawing the attention of the boardroom. Importance of cybercrimes can be realized from the experience of Australia where four out of five (80 percent) of Australian CEOs rate Cyber Security as a top investment priority and amongst the top five risk areas for their business (KMPG, 2018)

14.3 Examples of Cyber Supply Chain Crimes

Hackers have actively exploited several vulnerabilities in supply chains globally by leveraging many different forms of attacks, for example Trojan attacks and Distributed Denial of Service (DDoS) events, which are discussed below.

14.3.1 Hacking

In 2016, hackers hacked into a significant section of the internet through multiple coordinated DDoS attacks. Global DNS (Domain Name System) provider DynDNS responsible for aligning domain names to IP addresses was inundated with DNS lookup requests from hacked IoT devices such as webcams and printers. Major services such as AirBNB, Amazon, GitHub, Netflix, PayPal, and Twitter experienced multiple outages as a result of the event (KMPG, 2018).

14.3.2 Use of USB Infected with Virus

A third party managing the maintenance of speed cameras in a government department in Australia during 2016 used a USB inadvertently with the 'WannaCry' virus to make updates across devices. The Government was compelled to review the accuracy of all fines imposed during the period of infection passed on by negligence (KMPG, 2018).

14.3.3 Explosion of Notpetya

1n 2017, Notpetya exploded across the world affecting businesses including shipping ports, supermarkets, hospital services, advertising agencies, and law firms – even causing stoppage of operations at the Cadburys chocolate factory in Hobart (KMPG, 2018).

14.3.4 Security Flaws in Wi-Fi Devices

KMPG (2018) reported about a security flaw, which could hack into any Wi-Fi device. The Key Reinstallation Attack (known as KRACK) exploits replacing the Keys used to authenticate to any Wi-Fi device. The highjack can be performed from anywhere within the Wi-Fi's range and can steal any data transmitted to or from the device. The remedy lies in updating the firmware on each of the devices, a very time-consuming exercise sometimes taking businesses months or even years to complete (KMPG, 2018).

14.3.5. Counterfeit Threat

Armed Forces Communication and Electronics Association Cyber Committee (AFCEACC) (2012) brought out many cases of counterfeit threats that caused loss of millions of dollars to US business and the US Government. Some of the cases are reported here:

> *"The Justice Department in the US revealed that Federal authorities over the past five year have seized more than $143m worth of counterfeit Cisco hardware and labels in a coordinated operation that's netted more than 700 seizures and 30 felony convictions"*
>
> - AFCEACC, 2012

The FBI, Immigration and Customs Enforcement and Customs and Border Protection agencies worked to crack down on the bogus routers, switches, and other networking gear. In addition to costing Cisco and other US businesses millions of dollars, the frauds could threaten national security by infusing critical networks with gear that is unreliable or, worse, riddled with backdoors.

In another case, as part of the operation, a Saudi citizen residing in Texas, was sentenced to 51 months in prison and ordered to pay Cisco $119,400 after being found guilty of trying to sell counterfeit gear to the US Department of Défense. In 2008, the Saudi attempted to traffic one hundred gigabit interface converters that were bought in China and contained labels fraudulently indicating they were genuine Cisco equipment, according to court documents (AFCEACC, 2012).

In another case, a Chinese resident in the US was ordered to serve prison time and pay restitution of $790,683 for trafficking counterfeit Cisco gear. The prospect that government and business networks may have deployed bogus gear has raised national security concerns, since much of the counterfeit equipment originates in China (AFCEACC, 2012).

Since late 2007, US authorities have made more than 1,300 seizures of 5.6 million bogus semiconductors. More than 50 shipments were falsely marked as military or aerospace grade devices (AFCEACC, 2012).

14.3.6. Ransomware

The Colonial Pipeline is the largest pipeline for refined oil products in the U.S. having two tubes of 5,500 Miles (8,850 KM) long with carrying capacity of 3 million barrels of fuel per day between Texas and New York. The pipeline became the target of one of the most serious malwares cyberattacks during 2021 (on May 7, 2021) (Kelly, 2021). The ransomware cyberattacks adversely affected the computerized equipment of the Colonial Pipeline. The Colonial Pipeline Company stopped the operations of the pipeline to limit the effect of the cyberattack. The period of the stoppage continued for 5 days resulting in temporary fuel shortage along the East Coast. The cyberattack was the largest cyberattack on an oil infrastructure in the history of the U.S.A. (Gonazola et al., 2021). The Colonial Pipeline Company paid a ransom of 75 bitcoin (USD 4.4 million) to the hacker group to get a software application for restoration of the network operations. Later, the Department of Justice informed that it had recovered an amount of 63.7 Bitcoins ransom payment from the hackers (Mallni and Barr, 2021).

The attacks mentioned above are some of glaring examples of cyberattacks and their negative impacts on the society. Many cyberattacks are happening in various areas of activities of supply chain, which need to be controlled.

14.4. Cyber Supply Chain Risk Classification

Classification of cyber supply chain risks is done differently by different authors as perceived by them. Brown (2017) classified the Cyber Supply Chain (CSC) in four categories as IT, non-IT, management, or government. Ghadge et al. (2019) classified the cyber supply chain risks in five categories: physical threats, breakdown, indirect attacks, direct attacks, and insider attacks. Pandey et al. (2020) have categorized the cyber supply chain risks more rationally into three categories: supply risks, operational risks, and demand risks linking to the upstream and downstream SC.

14.4.1 CSC Risk Classification by Brown

According to Brown (2017), IT risks are technical and operational, while non-IT risks relate to organizations, products and services, environmental factors, and natural disasters (see Figure 14.1 above). Supply chain risks spread exponentially as an attack on one party may pass on to others due to interactions between the organizations that are integrated through processes.

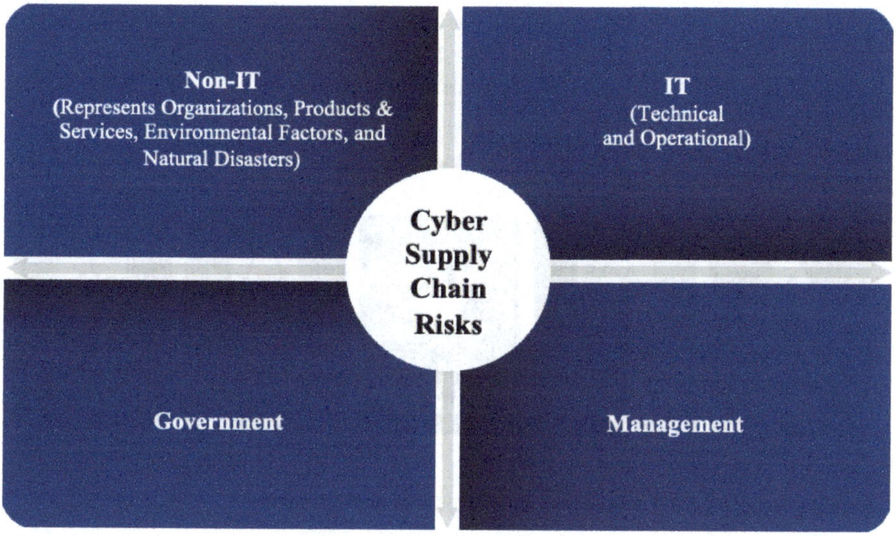

Source: Brown (2017)

Figure 14.1: CSC Risk Classification by Organizational Elements

14.4.2 CSC Risk Classification by Ghadge

Ghadge et al. (2019) classified cyber risks in five categories: Physical threats, breakdown, indirect attacks, direct attacks, and insider attacks (Figure 14.2).

i. **Physical Threats**: Threats to physical components and machines such as computers and printers is a physical risk caused from any event or action by someone. Natural disasters like floods, earthquake, accident, lightening, and terrorist attack can cause physical damage to the physical components and systems. Cyber-attacks damaging the physical systems and components are cyber risks (Urciuoli and Hintsa, 2017).

ii. **Insider Threats**: Employees of the company can be a major risk to the company's cyber security. Employees can compromise with the confidentiality of the passwords (Stephens and Valverde, 2013). Keeping the passwords in notes for quick reference and disclosing information during discussions with colleagues or others are some of the common actions by the company employees. Such actions of the employees, intentional or unintentional, can be the cause of threat to the company's cyber security (Kunnathur, 2015). The cyber risk is great when the employees manage access to company's information for personal gains (Sharma and Routroy, 2016).

Source: Ghadge et al (2017)

Figure 14.2: CSC Risk Classification by Type of Attack

iii. **Direct Attacks**: The cyber risk of deliberate action can be divided into two categories - direct attacks and indirect attacks. Direct attacks are carried out by malicious actions by someone outsider or insider for financial gains. Such acts include hacking attacks (Deane et al., 2009), denial-of-service (Faisal et al., 2007) or password sniffing (Warren and Hutchinson, 2000).

iv. **Indirect Attacks:** Attackers access the target system indirectly by viruses, worms, and Trojans (Jones and Horowitz, 2012). Counterfeit products, soft- and hardware (Boyes, 2015), malicious codes (Kunnathur, 2015) and spoofing attacks (Smith et al., 2007) also fall in this category. Verizon (2018) argues that cyber-attacks that originate via phishing, i.e., gaining access to sensitive information by disguising the threat as a trustworthy entity, are increasing these days.

v. **Breakdown:** The outdated firewalls and overdue security updates can cause breakdown of the systems and the resources (Tran et al., 2016). Breakdown failures are potential risks which are easier to predict than natural disasters or intentional attacks. But their impact can be equally severe.

The company-originated cyber risks form a major part of all the cyber risks faced by an enterprise. A survey conducted in 2018 by cyber security experts showed that 53% of the institutions experienced an internal attack. 51% of respondents were concerned about accidentally clicking on phishing links, while 47% were concerned about malicious employee behavior (AON 2019 Cyber Security Risk Report).

14.4.3 CSC Risk Classification by Pandey

Pandey et al. (2020) have categorized the cyber supply chain risks into three categories relating to the upstream and downstream SC, which include supply risk, operational risk, and demand risk as presented in Figure 14.3.

i. **Supply Risks**: Supply risks are caused by the suppliers or the supply market through inbound supplies. These risks make the focal firm unable to satisfy the customer demand. At times, suppliers do not supply good quality materials and customers are unable to find the cause of the problem. Inappropriate manufacturing standards of suppliers leads to question of authenticity of the supplied item. This needs to have a cyber risk management program. Examples of the supply risk can be seen from the activities of Dragonfly, the cyber-espionage group targeting the pharmaceutical manufacturing companies in North America and Europe since 2011 (Pandey et al., 2020). This group has targeted the companies through their SCs by trojan using the legitimate ICS software. This led to compromise of the genuine files of the depository by replacing them with files added with malware. The malware was then installed alongside the ICS software after downloading from the supplier's websites. The targeted company is not able to detect the problem of the compromised software due to changes made at the source.

ii. **Operational Risks**: Operational risks take place due to serious breach of access controls of plant's ICSs by the attackers causing malfunctioning of plants that lead to physical damage and disruption in the operations. IT systems have been used extensively in digital supply chain management to speed up the operations, remove the complexities and enhance the performance, but they add to the problems relating to information risks associated with data losses, loss of integrity and information in the manufacturing process management. There are increasingly more problems in the integrity of the data for maintenance and in data processing. Operational risks are more due to lack of SC coordination that impacts employee systems and procedures in SC (Bhattacharya et al., 2010). Failure of machinery facilities, design risks, and shortage of skilled labor are the main operational risks.

iii. **Demand Risks**: Failure of market/brand results in demand disruptions. In CPS, the supply-side risk related to safety issues, which impact the perceptions of the public and their associated demand, also generates the demand-side risk. Examples include problems with national authority's websites on the deadline day for submission of personal tax returns, the collapse of ticketing systems for major events such as concerts and sporting events and the launch

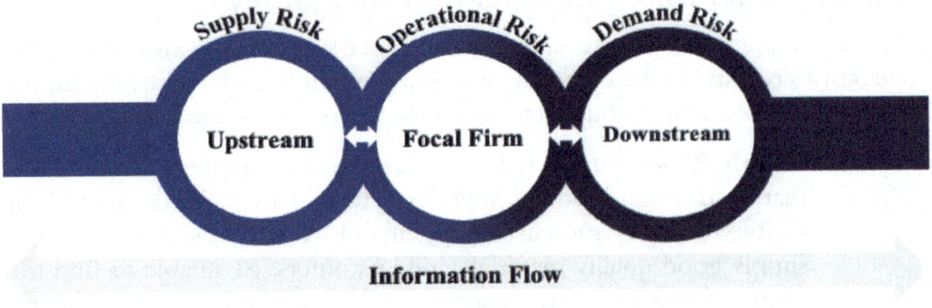

Source: Pandey et al. (2020)

Figure 14.3: CSC Risk Classification by Stage of Supply Chain

of online sales events (Boyes, 2015). Tummala and Schoenherr (2011) analyzed the sources of the demand risks as order fulfilment errors; inaccurate forecasts; information distortion. Mangla et al. (2016) analyzed competing risk as one of the demand risks, which are the tremendous risks faced by companies as the market is becoming competitive. The other category of demand risk elaborates that customers are important for implementing the successful practices of efficient SCM (Muduli et al.,2013). In the context of SC, it is important to have awareness about customer reliability to manage the demand related risks of the product (Mangla et al., 2016). Demand risks occur because of a lack of trust across SC companies. These are affected by false information relating to company websites, business licensees, trade documents, product test certificates, etc. For example, some companies attract customers illegally by stealing the identity, name of legitimate and well-known companies to perpetrate illicit operations.

Unexpected changes in the market result in demand disruptions because of errors in order fulfillment, mistakes in the forecasts, and distortion in the information. Competition in the market, customer attitude and changing preferences are other important factors influencing demand risks. Companies should, therefore, remain connected with the customers. Distrust in the company can significantly influence demand risks. Wrong and false information about the company, its website, licenses, trade documents, and the test certificates are some of the causes for the loss of trust in the company. Some fraudulent or unethical companies adopt the name or trademark that is very similar to the name of a reputed company to earn revenue by cheating the customer that is operating under the assumption of doing business with the more reputed company.

Table 14.2 below lists the different sources of the cyber supply chain risks as classified by Pandey et al (2020).

TABLE 14.2: SOURCE OF SUPPLY, OPERATIONAL, AND DEMAND RISKS

Supply Risks	Operational Risks	Demand Risks
Difficult accessibility of suppliers	Poor functioning of the plants	Stealing the intellectual property
Copying/theft of supplier Identifications	Operations of the plant interrupted suddenly	Tempering with the data
Unauthorized access to the network of the suppliers	Unable to find the coding error	Accessing the customer's data without authority
Addition of malware to make changes in source code	Product specification intentionally misguided	Intentional wrong communication
Supply of Compromised software	Stealing of data	Interfering with the information
-	-	Channels of payments without authority

14.5 Cyber Supply Chain Risk Mitigation

14.5.1 Risk Mitigation Measures

Measures for mitigating the cyber risks can be classified in three phases: pre-attack, trans-attack, and post-attack (Ghadge et al., 2019). Various measures for mitigating the cyberattacks under each phase are presented in Figure 14.4.

14.5.2. Risk Mitigation Strategies

Strategic plan to mitigate the cyber security risks is presented in Figure 10.5. Various cyber security attacks such as password attacks/cracking software, spoofing attacks, impairing the services, direct attacks, malicious activities, and the insider attacks impact the upstream SCs, focal firm, and downstream SCs.

The cyber risks are classified in three classes: supply risk relating to the upstream SCs, operational risks relating to focal firms, and demand related risks linked to downstream supply chains. These cyber supply chain risks can be mitigated by a strategic plan comprising three strategies as below (Pandey et al. (2020) (see Figure 14.5 on next page):

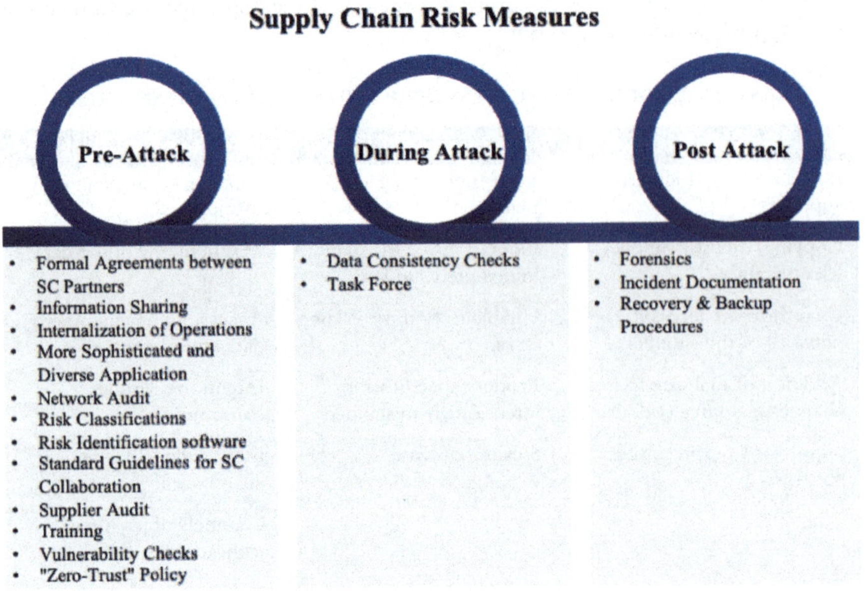

Source: Ghadge et al. (2019)

Figure 14.4: Risk Measures by Stage of Supply Chain Attack

i. **Software Reliability**: Software reliability ensures that there are no vulnerabilities, and the system should behave as planned. Simpson et al. (2010) developed the software based on the principle of the system reliability with minimum risks in global SC. The reliability-based model had three features: Security, integrity, authenticity. The model should be able to anticipate and address any vulnerability to offer the desired security. The software during its sourcing, creation, and delivering to the customer should be capable of removing any vulnerability and demonstrating its integrity. Further, the customer getting the software should be convinced that the software is original and authentic and is not a counterfeit product.

ii. **Data Management**: Data management system of the company should be designed and managed in such a way that data does not get leaked and should be correct. For example, data management in the case of a manufacturing partner is not leaked by the supplier to any other agency or manufacturer. Collaborations between the manufacturer and supplier or between one manufacturer with another manufacturer should be facilitated to share data between the partners easily, and safely. In a competitive business environment, the collaborative partners are concerned about the possibilities of data leakage. Data management system should have safeguards,

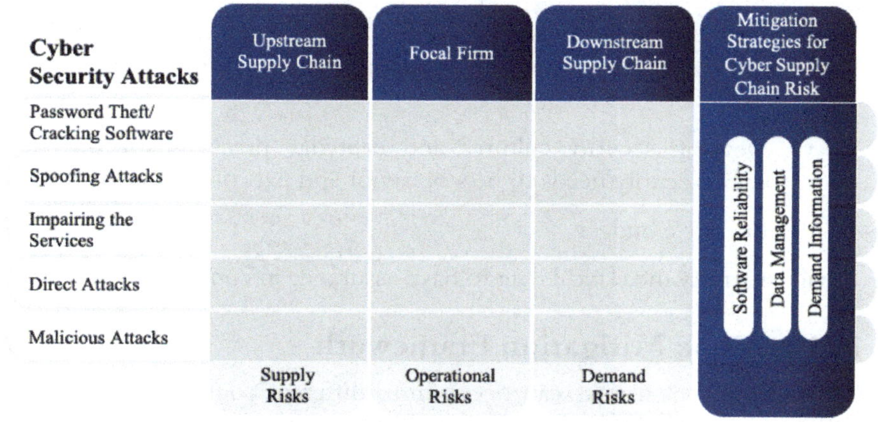

Figure 14.5: Strategic Plan to Mitigate Cyber Supply Chain Risks

firewalls, and adequate training on various issues so that it is not possible to leak out any confidential and unauthorized information to the collaborative partners.

iii. **Information for Demand Management**: Demand related information should be shared easily between the partners. The sharing of information on demand is essential for managing the operations effectively in a timely manner. The use of Cloud technology used in sharing of information, however, creates the possibilities of cyber risks. The information sharing system should be robust enough to protect the information shared and simultaneously offer timely, correctly, and safely the information between the partners. Use of standards, guidelines, and best practices help in mitigating the cyber security risks in supply chain management.

14.5.3. Risk Mitigation Plan

Following strategic plan can help control the cyber-attacks on supply chain of an organization:

i. The first step in controlling the cyber-attacks is to identify the most sensitive data to be protected and prioritized. Decision support methodologies and tools can help quantify and rank cyber risk.

ii. Next step is to identify the 'Threat Actor' and the way these 'Threat Actors' likely to attack.

iii. Evaluate the controls in place and the gaps to be addressed. Identify the level of acceptable risk.

iv. Prepare a plan 'prevent, detect, and respond with a defined and tested response plan, and be prepared to meet the challenge of an attack.

v. Develop security cultures and everyday practices for the staff. Cybersecurity needs to be a behavior and part of the business.

vi. Prioritize budget

vii. Identify the critical data to have assurance in controls.

14.5.4. Risk Mitigation Framework

Cyber supply chain risks can occur from different sources. Ghadge et al. (2019) proposed a 'supply chain cyber security system' as a unifying conceptual model (Figure 14.6).

The sources of cyber supply chain risks are either associated with IT (e.g., such as a direct or indirect attack), organizational (e.g., insider threat) or the supply chain (e.g., physical threat) systems. Ghadge et al. (2019) state that all three diverse elements namely, IT system, organization process, and supply chain security system (which includes process in supply chains infrastructure network) should be aligned to manage cyber risk. IT security systems can take care of cyber threats with the help of hardware, the latest technology and secure software platforms. Organizational security systems mitigate cyber-attack by physical assets, adhering to a set of guidelines and by raising awareness among employees.

Information sharing, collaborative risk management, and adaptability are the key strategies for supply chain security. This interlinked relationship between different (sub) systems is shown in overlapping circles in Figure 14.6. The distinct mitigation strategies, shown in the triangles, are critical for managing cyber risk in supply chains. Coordination of these security systems, joint information sharing and applying appropriate mitigating strategies can effectively manage cyber risk in supply chains.

14.6. Cyber Supply Chain Risk Management

COVID-19 pandemic has demonstrated over the last two years what can go wrong when a prolonged disruption causes sudden swings to supply and demand.

From the stockout of commodity items like paper products, canned goods, face masks, etc. to supply shortages and factory shutdowns, many businesses have endured what is perhaps the most challenging single event in the collective memory. As businesses continue working toward a normal state of

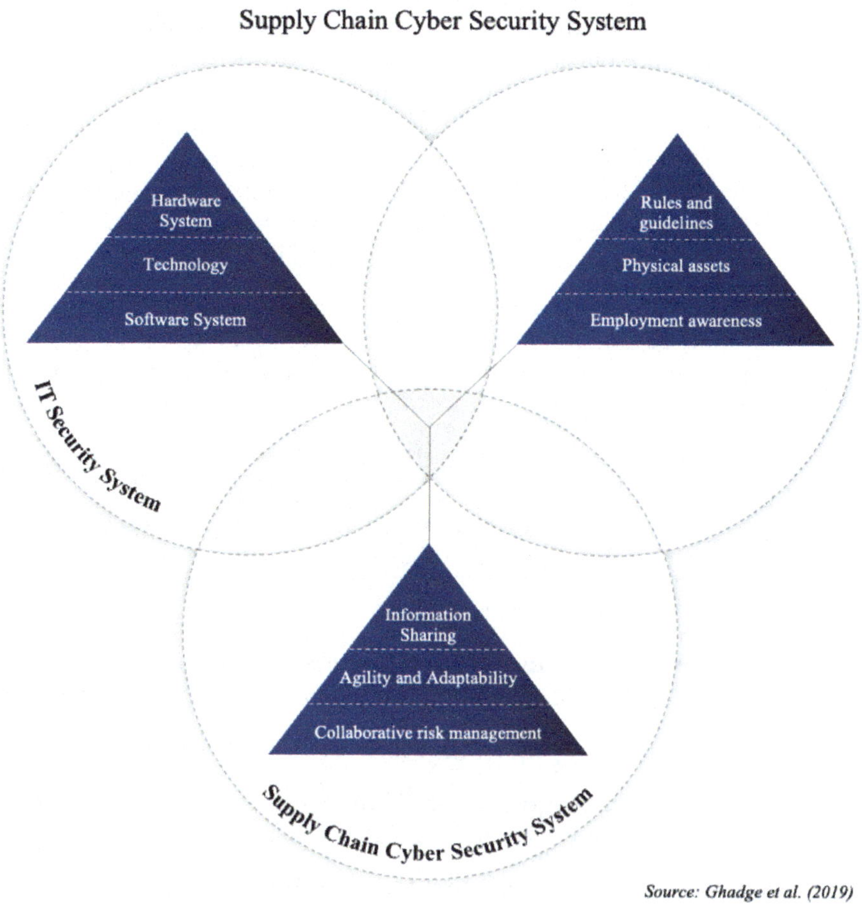

Figure 14.6: Cyber Supply Chain Risk Mitigation Framework

operations, the lessons learned from COVID-19 are prompting business leaders to reassess and rethink their approach to overall risks for the organization – especially across the supply chain, where preparation can go a long way to driving the appropriate response to a disruptive event.

This section describes a Cyber Supply Chain Risk Management process that should be leveraged by organizations globally to understand, plan, and prepare for any risk scenario leading to the creation of a flexible and resilient supply chain that can weather unforeseen disruptions from natural disasters, pandemics, etc.

The Cyber SC Risk Management process should involve five major steps as shown in Figure 14.7 on the next page:

i. **Risk Identification**: This is accomplished by drilling down both within and outside the organization to specifically identify the elements that can present a potential risk to the Supply Chain of an organization. Ideally, this should be executed by creating a specialized Supply Chain risk taskforce whose members possess the specific skillsets and knowledge of organizations Supply Chain resources, processes, and technology solutions to thoroughly, and efficiently, identify risks of each type and level.

ii. **Risk Evaluation**: Once a set of risks has been identified, these risks should be classified into risk categories using one or more of the approaches presented in Section 14.2 above. Key factors to include in such an analysis are:

- Probability of the risk occurring, including the identification of vulnerabilities in the systems, determination of the intention of malicious actors, etc.
- The relative impact of the risk to cost, revenue, profitability, brand, etc.
- The relative ability of the organization to mitigate the risk, including identification of alternate sources of supply, etc.
- The urgency with which the risk will have to be mitigated
- The degree to which the risk is acceptable in terms of political, environmental, social, economic, etc.
- The potential for the risk to grow in magnitude and/or impact over time, and hence the urgency to act

iii. **Risk Planning**: Having identified and assessed risks to Supply Chain due to natural disasters, pandemics etc., a detailed Risk Plan should be created that identifies the following at a minimum:

- Mitigation approach to use (accept, avoid, transfer, or reduce the risk)
- Mitigation strategy to use (such as software reliability, data management, or information related to demand management etc. as explained in section 14.5.2 above)
- Mitigation plan to use (such as the one explained in section 14.5.3 above)

Figure 14.7: Cyber Supply Chain Risk Management

- Resources that will be responsible for executing different steps of the plan along with the estimated time, sequence of remediation activities, dependency between activities, and the approvals that will be needed in each scenario, etc.

iv. **Risk Monitoring**: Since risks, by definition, are often unexpected and unpredictable, the cyber supply chain risk management process must define systematic reporting and monitoring of all ongoing and uncovered risks. This can, and should, include the standard reporting systems to be leveraged, the reports that should be generated along with the frequency of such report generation, report distribution, and key measures to be reported on (as described in Section 14.5.1 above.

v. **Risk Control**: The goal of such a Supply Chain Risk Management process is not to eliminate all risks from the supply chain as that would be impossible to achieve and financially not feasible in most cases. However, if such a process drives greater flexibility and responsiveness in the supply chain of an organization thereby making it more resilient than before, the Risk Management process would be considered successful from most perspectives.

14.7 Summary

This Chapter focuses on cyber risks in the supply chain. Different authors have defined the cyber supply chain risks differently over the years. Definitions are discussed differently from time to time, details are given right from the year 1987 to 2018. An updated definition has been developed and presented in the Chapter. Similarly, cyber supply chain risks have been classified differently by Ghadge et al. (2019), brown (2017), and Pandey et al. (2020). Pandey et al. (2020) classified the cyber security risks across the upstream and downstream in three categories: supply risks, operational risks, and demand risks. The Chapter highlights the sources of supply risks, operational risks, and demand risks. Digital technologies implemented in 'Industry 4' have been found contributing to the supply chain risks.

Cyber supply chain risks have been explained by examples. Brief details of the Ransomware Cyberattack on Colonial Pipeline have been discussed with its adverse impacts on the supply of fuel to transport systems along the East Coast in the USA. Fuel shortage continued for 5 days and was restored only after payment of bitcoins (USD 4.4 million) to the hackers by the affected company. Later, Department of Justice recovered from the hacker's bitcoin 63.5, part of the ransom paid to them.

Chapter discusses the measures to mitigate the cyber risks under three phases: pre-attack, trans-attack, and post-attack phases. This is followed by a strategic plan to control the cyberattack and framework for mitigating supply chain risks.

References

AFCEACC (2012). Supply Chain Risk Management Awareness. Pdf 1-13.

AON Cyber Security Risk Report (2019). Retrieved from https://www.aon.com/getmedia/4c27b255-c1d0-412f-b861-34c5cc14e604/Aon_2019-Cyber-Security-Risk-Report.aspx

Bhattacharyya, K., Datta, P. and Offodile, O.F. (2010), "The contribution of third-party indices in assessing global operational risks", *Journal of Supply Chain Management*. 46 (4). 25-43.

Boyes, H. (2015), "Cybersecurity and cyber-resilient supply chains", *Technology Innovation Management Review*. 5 (4). 28.

Brown, D.A. Best Practices in Cyber Supply Chain Risk Management; Intel; 2017. Available online: https: //www.nist.gov/document-18221. (Accessed on 20 October 2018).

Cebula, J.L. and Young, L.R. (2010), A Taxonomy of Operational Cyber Security Risks (No. CMU/SEI-2010-TN-028). *Carnegie-Mellon University Pittsburgh PA Software Engineering Inst*, Pittsburgh PA.

Christopher and Kelly Stephanie. (2021). Cyber attack shuts down top U.S. fuel pipeline network. *Reuters*, May 8, 2021.

Deane, J. K., Ragsdale, C. T., Rakes, T. R. and Rees, L. P. (2009). Managing supply chain risk and disruption from IT security incidents. *Operations Management Research* 2 (1-4). 4-12.

Faisal, M. N., Banwet, D. K. and Shankar, R. (2007). Information risks management in supply chains: an assessment and mitigation framework. *Journal of Enterprise Information* Management. 20 (6). 677-699.

Ghadge, A., Wurtmann, H., & Seuring, S. (2019). Managing climate change risks in global supply chains: a review and research agenda. *International Journal of Production Research.* 1-21.

Gonzalez, Gloria; Lefebvre, Ben; Geller, Eric (May 8, 2021). Jugular' of the U.S. fuel pipeline system shuts down after cyberattack. *Politico.*

Jones, R. A. and Horowitz, B. (2012). A system-aware cyber security architecture. *Systems Engineering.* 15 (2). 225-240.

Jüttner, U., Peck, H. and Christopher, M. (2003). Supply chain risk management: outlining an agenda for future research. *International Journal of Logistics Research and Applications.* 6 (4). 197-210.

KPMG. (2018). Digital Supply-Chains-the hype and the risks. *KPMG International.* 1-8.

Kumar, S.K., Tiwari, M.K. and Babiceanu, R.F. (2010). Minimization of supply chain cost with embedded risk using computational intelligence approaches. *International Journal of Production Research.* 48 (13). 3717-3739.

Kunnathur, A. (2015). Information security in supply chains: A management control perspective. *Information and Computer Security*, 23 (5). 476-496.

Mallin, Alexander and Barr, Luke. (2016). Critical success factors for reverse logistics in Indian industries: a structural model, *Journal of Cleaner Production.* 129. 608-621.

March, J.G. and Shapira, Z. (1987). Managerial perspectives on risk and risk taking. *Management Science.* 33 (11). 1404-1418.

Muduli, K., Govindan, K., Barve, A. and Geng, Y. (2013). Barriers to green supply chain management in Indian mining industries: a graph theoretic approach. *Journal of Cleaner Production.* 47. 335-344.

Pandey, Shirpa. (2020). Cybersecurity risks in globalized supply chains: conceptual framework. *Journal of Global Operations and Strategic Sourcing.* 1-27.

Sharma, S. and Routroy, S. (2016). Modelling information risk in supply chain using Bayesian networks. *Journal of Enterprise Information Management.* 29 (2). 238-254.

Simpson, S., Baldini, D., Bits, G., Dillard, D., Fagan, C., Minnis, B. and Reddy, D. (2010). Software Integrity Controls – an Assurance-Based Approach to Minimizing Risks in the Software Supply chain, SAFECode, available at: https://safecode.org/publication/SAFECode_Software_Integrity_Controls0610.pdf (accessed 19 December 2019).

Stephens, J. and Valverde, R. (2013). Security of E-Procurement Transactions in Supply Chain Reengineering. *Computer and Information Science.* 6 (3). 1-20.

Tummala, R. and Schoenherr, T. (2011). Assessing and managing risks using the supply chain risk management process (SCRMP). *Supply Chain Management: An International Journal.* 16 (6). 474-483.

Tran, T., Childerhouse, P., Deakins, E. (2016). Supply chain information sharing: challenges and risk mitigation strategies. Journal of Manufacturing Technology Management. 27 (8). 1102-1126.

Urciuoli, L. and Hintsa, J. (2017). Adapting supply chain management strategies to security - an analysis of existing gaps and recommendations for improvement. *International Journal of Logistics.* 20 (3). 276-295.

Verizon (2014). Data breach investigations report. verizon.com. 13-15.

Verizon (2018). Data breach investigations report, Available at: http://www.verizonenterprise.com/industry/public_sector/docs/2018_dbir_public_sector.pdf ((Accessed: 18.07.2019).

Wallace, M. (2016). Mitigating Cyber Risks in IT Supply Chain. *The Global Business Law Review.* Cleveland-Marshal College of Law. Library. Cleveland State University.

Yeboah-Ofori, A. and Islam, S. (2019). Cyber security threat modeling for supply chain organizational environments. *Future Internet.* 11 (3). 63.

Section IV

Supply Chain Management: The Road Ahead

15

Strategies to Manage Major Disruptions in SCM

Supply Chain Management (SCM) is the lifeline of our daily life. We get our daily requirements such as food, electricity, water, gas, medicines through supply chains. All the supply chains are usually designed to take care supply and demand in a balanced manner. However, every day, there are some disruptions in supply chains. Minor disruptions can be managed by us without much problem by little change in our schedule or by using some additional resources. High degree of disruption may be due a disaster. A major disruption happens due to a major disaster. The gravity of the disaster decides the level of disruption which can be low, medium, high, or very high. In case of major disruptions/disasters, supply chains are seriously affected and are not able to meet even the demand of the daily needs.

15.1 Disaster

Disaster can be due to natural calamities such as heavy rains flooding an area, earthquake, cyclone, jungle fire, volcano abruption or due to man-supported activities like rail/airplane/bus accident, explosion, gas leakage from an industrial plant, fire, and stampede. In today's time, disasters can be foreseen with the help of advanced technological solutions,

According to Ivanov and Wendler (2019), "disaster" or "catastrophe" is a serious disruption of the functioning of society, disturbing its priorities, health, life, or property of the environment. The disruption may be the result of nature or human activity that has affected at least 100 people with injuries and killed 10 people (Van Wassenhove, 2006; Kovacs and Spens, 2009). The impact may be in physical, social, economic, ecological aspects, and the general condition of the population concerned, thus influencing the population's poverty, and consequently its vulnerability. Based on the speed and trigger, disasters can be classified as having slow- or sudden onset, and natural or man-made causes (Van Wassenhove, 2006).

A disaster is a serious disruption of the day-to-day life and activities of a community, and society (ADPC, 2019). It disrupts the preexisting arrangement between demand and supply. Disaster leads to widespread loss of human life, materials, property both private and public, environmental losses, injury, and diseases. It leads to destruction of assets, loss of services, social and economic disruption, and environmental degradation. Disaster has negative impacts on physical and mental health of the people. The affected people, community or society are not able to stand the impact of the disaster with the available resources and need help from government and other non-governmental agencies.

15.1.1 Examples of Major Disasters

Major disasters can happen at any part of the world, but some areas are more prone to disasters due to cyclones, earthquakes, jungle fires, heavy rains/flooding. Repeat cyclones, earthquakes, flooding of the area and forest fires happen in these areas interrupting the normal life of the people of the area. Some cases of major disasters and their impacts on the human life and SCM are presented here.

Disaster 1: Flooding in Somalia in 1997 (Ivanov & Wendler, 2019)

Impact of the Disaster:

i. Shabelle and Juba rivers in southern Somalia overflowed due to rain.
ii. Estimated 2 000 deaths
iii. Left 250 000 people without shelter.
iv. Severe damage to infrastructures and agriculture
v. Crop losses up to 80 % in the most affected regions until 1999 (Lovgren and English, 1997; Menkhaus, 2010).
vi. About 1 million people were affected by food shortages and risk of starvation.

Disaster 2: Cyclone Nargis in Myanmar, Burma in 2008 (Ivanov & Wendler, 2019)

Impact of the Disaster:

i. Official count of people dead and missing was more than 60000
ii. The UN even estimated around 100 000 people to have been injured or killed.
iii. Housing of more than one million Burmese was destroyed, leading to internal displacement of 800 000 people.

vii. Hundreds of thousands remained missing. (Liu and Seaton, 2008; Ratnesar, 2008).
viii. Storm reached peak winds of 215 km/h (storm of category 4)
ix. Wind and tidal waves caused vast devastations in Sri Lanka and five regions of Myanmar (Paik, 2011).
x. The cyclone destroyed villages, flattened crops, flooded rice fields with salt water and killed thousands of farm animals (Seth, 2008)
xi. Food reserves were destroyed and market prices for main foodstuff rose, further fueling the food insecurity of some population groups (Southeast Asia, 2008; New Scientist, 2008).

Disaster 3: Earthquake in Haiti in 2010 (Ivanov & Wendler, 2019)
Impact of the Disaster:

i. January 2010 at 4.53 p.m. local time, an earthquake of magnitude of 7,2 devastated the region around the capital of Haiti, Port-au-Prince.
ii. Caused 220 000 dead and 300 000 injured.
iii. Two million Haitians were left without shelter.
iv. The losses in infrastructure and housing, rose to USD 7.8bn (FAO, EU, 2010)

Disaster 4: The Triple disaster in Japan in 2011 (Nagurney, 2017)
Impact of the Disaster:

i. Happened on March 11, 2011.
ii. More than 8000 deaths in the Fukushina area alone.
iii. Disrupted the high tech, automotive, and even food industries.

Disaster 5: Haiyan Typhoon in Philippines in 2013 (Nagurney, 2017)
Impact of the Disaster:

i. A very powerful tropical cyclone that devastated portions of Southeast Asia, especially the Philippines, on November 8, 2013.
ii. Deadliest typhoon
iii. Killed at least 6,190 people in that country alone.
iv. As of January 2014, bodies were still being found.
v. The overall economic losses totaled $10 billion.

Disaster 6: Ebola in Liberia, Guinea, Sierra, Leone, Nigeria, the US, and Mali in 2014 (Nagurney, 2017)
Impact of the Disaster:

i. First case recorded on March 23, 2014.
ii. At least more than 11,178 people died from Ebola in six countries.
iii. Total number of recorded cases was more than 27,275.

Disaster 7: Earthquake in Nepal in 2015 (https://www.britannica.com/topic/Nepal-earthquake-of-2015)

Impact of the Disaster:

i. The 7.8 magnitude earthquake on April 25, 2015, and its aftershocks.
ii. Killed nearly 9,000 people.
iii. Injured 22,000 others.
iv. The disaster pushed about 700,000 people below the poverty line in the Himalayan nation.
v. About 500,000 homes were made unlivable.
vi. Left about three million people homeless.
vii. According to The Wall Street Journal, Nepal needs $6.66 billion to rebuild.

Disaster 8: Harvey wide-spread flooding around Houston, Texas, USA in 2017 (FEMA, 2017)

Impact of the Disaster:

i. The flooding of the large number of housing societies including roads in Houston and surrounding areas occurred due to unprecedented rains. Many relief camps were created in school buildings and other places. Large number of people took shelter in the relief camps.

Disaster 9: Irma in South Florida, USA in 2017 (FEMA,2017)

Impact of the Disaster:

i. Largest evacuation carried out in U.S. history.

Disaster 10: Maria devastating the power grid in Puerto Rico, USA in 2017 (FEMA, 2017)

Impact of the Disaster:

i. Left most of the island without power for almost three months
ii. Delaying full power restoration until late 2018.
iii. Combined impact of the above three hurricanes in USA in 2017 (FEMA, 2017)
iv. The three hurricanes Harvey, Irma, and Maria caused a combined $265 billion in damage, and each ranked among the top five costliest hurricanes on record.
v. The three storms in Disaster 8, 9, and 10 impacted more than 48 million people
vi. Prompted 5 million homes to register for government assistance.
vii. FEMA used over 17,000 personnel to help people
viii. FEMA provided survivors with more than $2 billion in essential commodities during the 2017 hurricane season.

ix. FEMA Urban Search and Rescue Task Forces, composed of state and local emergency responders, saved, or assisted nearly 9,500 people across the three hurricanes.

15.1.2 The Priority after Disaster

Immediate priority after the disaster is to restore the preexisting supply chains to bring normalcy in supply and distribution of essential products like milk, water, electricity, gas, pharmaceuticals, medicines, medical goods, fuels, and other products. Repair and restoration of the essential infrastructures like electric supply systems, water supply systems, gas installations, rail stations, bus stations, and hospitals are the next thing to be taken care of. Further, in case of highly populated areas, attempts should be to restore normal life at the earliest. Reopening of the schools, colleges and universities is important to keep the younger generation occupied productively and to avoid any disturbances including the law-and-order situation which should always remain under control. Restoring the transport system and transport network is essential for ensuring supply of daily use items in time, and at the right place. Efficient communication system is necessary for quickly bringing the order quickly and delivering the right information at the right place and at the right time. Delays in delivery of essential products leads to uncertainties in the consumers and escalates the demand for products. Effective management of disaster is required to keep disruption under control.

15.1.3 Disaster Risk

Disaster risk is related to the severity of the disaster. Disaster risk comprises different types of losses which cannot be easily quantified although they can be assessed and mapped with reasonable accuracy.

15.2 Disaster Management

Disasters of any magnitude create emergency needs for immediate action. Natural disasters mostly happen in a short time and call for immediate attention. Disaster management cycle shown in Figure 15.1 below has four components:

i. Mitigation

ii. Preparation

iii. Response

iv. Recovery

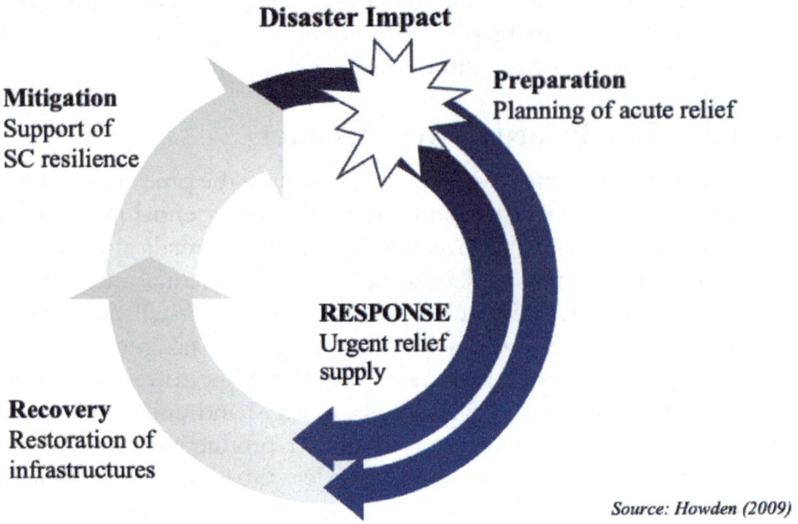

Figure 15.1: Disaster Management Cycle

15.2.1 Mitigation

Local governments take every action to reduce the sufferings of the population from the disasters and strengthen resilience of infrastructures. Prompt and advance actions taken by the local governments mitigate the severity of the impact of the disaster. Local farmers can help greatly in mitigating the shortage of food items. Field actors should identify the potential resources and establish contacts with them. They should assess their capabilities and capacities. They must study the bottlenecks in the existing setups and try to remove them beforehand. Risk and supplier analysis should be completed in advance to develop suitable strategy. Attempts should be made to establish contacts with local governments to improve resilience of the supply chains. Attempts should also be made to get aid from international aid organizations to tackle the issue of poverty and vulnerability.

15.2.2 Preparation

Preparation includes all actions taken in developing a strategy to provide prompt and effective disaster response.

15.2.3 Response

The Response refers to the actions taken immediately after the occurrence of the disaster to effectively establish temporary supply infrastructures and restoration of basic services to supply relief goods promptly to the affected population. Aid supply, and preparation of staff and local populations, are important aspects in responding promptly. Developing "Soft" skills, such as cultural awareness helps in improving the distribution channels. An integrated strategy should be developed for sustainable disaster relief through local supply sources and infrastructures already available.

15.2.4 Recovery

Recovery refers to actions taken to ensure sustainability and efficiency of reconstruction works, and rehabilitation of infrastructures and society. Sustainability of relief measures and restoration of existing infrastructure continues to be crucial for overall success of disruption management. Performance measurement is critical for continuous improvement. Regular (real time) monitoring and evaluation of relief actions needs to be channelized. New technologies and information systems may provide an important potential for improvement at all stages of disaster relief.

15.3 Actors and Stakeholders of Disasters

Different stakeholders and actors come forward to help those who are affected by the disaster. These stakeholders are from private and governmental organizations, willing to offer humanitarian aid. International organizations such as the World Food Program (WFP) helps people in need of food. The European Union (EU) or the North Atlantic Treaty Organization (NATO), facilitate the interaction with the authorities in the affected regions (Ivanov and Wendler, 2019). The military, the media, and private service providers also contribute to disaster relief. Coordination of the efforts of different Actors/Stakeholders is a challenging task in the light of their different positions and working principles. Their involvement in the relief work should, therefore, be organized in such a way that does not contradict with their principles and mandates.

15.4 Enhancing Resilience in Supply Chain

The resilient supply chain is an important tool in managing the disaster immediately. The disaster managers should contribute to the resilience of the supply chains in the following two ways (DHS, 2019):

i. To collaborate with supply chain partners that makes the supply chains of the critical commodities and the essential services more resilient.

ii. To develop an awareness campaign with the help of response and the recovery plans to inform about various vulnerabilities through information systems. Public gets news about the problem, the dimensions of the actions taken to contain the damage, the Do's, and the don'ts along with suggested recovery plans.

A good strategy adopted by the disaster managers is to bring together the public sector and private sector to collaborate and meet the emergency. The collaboration would improve the resilience of the supply chains. However, involving the private sector in the disaster management efforts must be done in a guarded manner, especially in cases, where legal issues are involved, or regulatory guidelines must be followed. The disaster managers must have awareness and understanding of the distribution system of the supply chains of that area or zone to make the supply chains more resilient. Improved awareness and in-depth knowledge of the distribution system will enrich the response and recovery operation plans making the supply chains more resilient.

Knowledge of legal parameters is essential in making the supply chains resilient, which may include government directions about driving license, temporary closure of certain routes due to safety reasons or some construction activities in progress, storage of sensitive and explosive materials, gases, and restrictions on certain procurement activities. In-depth knowledge of such aspects will help in making the supply chains resilient.

15.4.1 Community Lifelines

Response to a disaster is best managed jointly by the different stakeholders comprising governments, the private sector, and non-governmental organizations (NGOs) (DHS, 2019).

The community lifelines construct helps in responding to disaster promptly with unified purpose and effective communication between all the members of the community, which includes Federal, state, local, tribal, and territorial governments, and private sector and non-governmental entities. The joint effort facilitates in prioritizing, sequencing, and concentrating the response efforts in maintaining or restoring the essential services and infrastructure. DHS (2019) suggests seven community lifelines as shown in Figure 15.2.

The above lifelines offer opportunities to understand and examine complex supply chains for cross-sector coordination. Community lifelines also identify critical focus areas in solving secondary supply chain problems. For example, a stoppage of the power grid can lead to a rise in generator and fuel demand. Local gas prices increase due to the increased demand and/or a shortage in generator supply.

15.4.2 Key Considerations

Emergency Manager likes to examine the known and potential hazards in the supply chain, alternate routes, alternate transportation, and known restrictions in storage, handling, and/or transportation (DHS, 2019). The analysis may cover the supply chain nodes as well as related infrastructure. These are discussed here.

i. **Supply Source**: Supply source may be in another country, within the country, or in a difficult area. Which factor may have an impact on the supply from the source such as extreme weather conditions, natural disasters, political disturbance, or national holidays?

ii. **Distribution Points**: Single distribution points may not be functional due to some problem. Emergency managers should identify alternate distribution point(s) to make supplies by rerouting the supply from other distribution centers. It should be established if the distribution point needs support from the community lifeline.

iii. **Inventory**: Emergency manager, during logistics planning, must examine if the jurisdiction needs support from a community lifeline. Review should cover the standard inventory supply, re-supply interval, storage facilities, power requirements and Just-In-Time supply option during the emergency.

iv. **Assess and Re-entry**: Route security and escort needs are important issues and should be part of the partnership agreements and memorandum of understanding (MOU). Security during emergency events should be under the control of local police and military forces. The communication at the point of distribution should be strengthened to control access to the restricted areas. Personnel posted at the checkpoints, at barriers, and/or at the facility should know the contents in the shipment and at the destination. Procedures for clearance, checkpoint access, and access to the facility should be circulated to the concerned staff.

v. **Fuel**: Different vehicles use different fuels other than the common fuels such as diesel, gasoline, and flex-fuel. Person(s) should be identified in advance to receive the emergency fuel during and after the disaster. Procedures for fuel access, stockpile and re-supply should be available for smooth recovery of supply chains.

vi. **Routes**: Supply chain logistic plans should identify the main supply routes and alternative routes should be clearly identified. In case of a disaster, special considerations about weight and noise restrictions on routes may be relaxed in addition to other restrictions about the height for roads under bridges, and hazardous or flammable

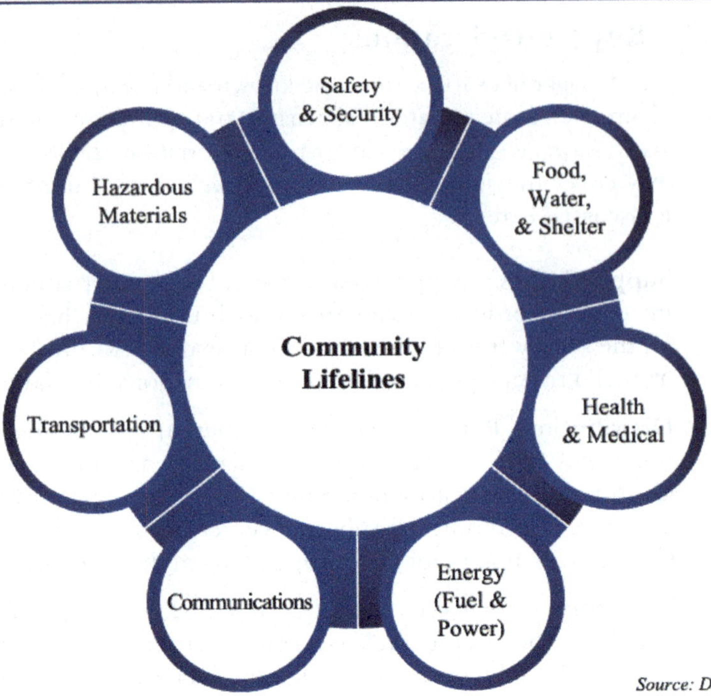

Source: DHS (2019)

Figure 15.2: Community Lifelines in Efficient Disaster Management

materials. Debris management plans should be available indicating who will clear the roads, priority of roads to be cleared first, and the destination for throwing the debris.

vii. **Transport Service Providers**: Different supply chains may need separate guidelines for the authorization and availability of pilots, drivers, and captains due to specific considerations. For example, in healthcare supply chains, there are strict rules permitting authority to transport certain products and the tracking and accountability of these products. In addition, there are restrictions on the transporters about the temperature and environmental controls in providing the service. Transportation operators control the movements and deliveries of their vehicles considering the weather forecasts.

Usually, there is a shortage of transport staff, drivers, and dispatchers during the disaster. Union member operators and volunteer operators may have different regulations for operation in comparison to other private or public operators. MOUs must be established in advance with unions, volunteer organizations, and private/public sector entities to avoid shortage of authorized drivers in an emergency.

Transportation operators may need authorization to transport materials and/or enter facilities. Personnel entering secure areas of maritime facilities and vessels require the Transportation Worker Identification Credential (TWIC) 12. Similarly, fuel truck drivers require site-specific certifications before they can receive fuel at individual distribution terminals.

15.5 Summary

This chapter is one of the most important chapters for supply chain management. Disasters bring disruption in daily life, business and functioning of the government. The definition of the disaster is explained along with the disruptions it brings. Ten selected cases of important disaster in different countries have been discussed bringing out the impact of the disaster in each case. The chapter discusses the immediate priorities after the disaster.

Chapter discusses disaster management with the help of the disaster management cycle, which comprises four elements: mitigation, preparation, response, and recovery. It identifies the actors and stakeholders of supply chains in the most testing times of disaster. Discussion then moves to strategy in enhancing the resilience of the supply chains necessary for managing the disruptions. Chapter identifies seven community lifelines that help in effective management of the disasters. The seven lifelines include safety and security, food, water, Shelter; health and medical; energy (power & fuel); communications; transportation; and hazardous materials. Chapter brings out the aspects that can strengthen the seven lifelines and help in managing the disasters efficiently.

References

ADPC (2019). Training Manual Disaster Risk Management (DRM) for Pakistan. *Asian Disaster Preparedness Center (ADPC)*. Bangkok, Thailand.1-196.

DHS. (2019). Supply Chain Resilience Guide. *Publication of Department of Homeland Security (DHS)*, USA. 1-44.

FAO, EU (2010). Special Brief: Haiti Earthquake, Food Security Information for Decision Making. *FAO*; online: http://www.fao.org/3/a-al053e.pdf [accessed: 7 July 2014].

FEMA. (2017). Hurricane season FEMA after-action report. Technical report 2018. *Federal Emergency Management Agency.*

Howden, M. (2009). How Humanitarian Logistics Information Systems Can Improve Humanitarian Supply Chains: A View from the Field, in: Lindgren, L., Jul,

S. (Eds.), Proceedings of the 6th International ISCRAM Conference – Gothenburg, Sweden, May 2009. Gothenburg.

https://www.britannica.com/topic/Nepal-earthquake-of-2015

Ivanov, D. and Wendler, E. (2019). Natural Disasters and Supply Chain Disruption Management. *ResearchGate.* 1-30.

Kovacs, G. and Spens, K. (2009). Identifying challenges in humanitarian logistics. *International Journal of Physical Distribution and Logistics Management.* 39. 506–528.

Liu, M. and Seaton, J. (2008). A Curse From the Heavens. *Newsweek* 151, 36–39.

Lovgren, S. and English, C.W. (1997). Floods could bring famine. *US News World Rep.* 123, 53.

Menkhaus, K., 2010. Stabilization and humanitarian access in a collapsed state: the Somali case. *Disasters.* 34. 320–341.

Nagurney, A. (2017). Topic 6: Disaster Relief Supply Chain SCH-MGMT 825 Management Science Seminar 'Advances in Variational Inequalities, Game Theory, and Applications, Spring 2017. *University of Massachusetts Amherst.* 1-102.

New Scientist (2008). Burma crisis holds lessons for us all. *New Scientist Editorial.*198, 5.

Paik, W. (2011). Authoritarianism and humanitarian aid: regime stability and external relief in China and Myanmar. *Pacific Review.* 24, 439-462.

Ratnesar, R. (2008). An Offer They Can't Refuse. *Time International* (Canada Edition). 171(21). 19.

Seth, A. (2008). Even Paranoids Have Enemies: Cyclone Nargis and Myanmar's Fears of Invasion. *Contemporary Southeast Asia* 30, 379–402.

Southeast Asia. (2008). Cyclone Turns Up Pressure on Regime. *Southeast Asia* 1. 7.

Van Wassenhove, L. (2006). Humanitarian aid logistics: supply chain management in high gear. *Journal of the Operational Research Society.* 475–489.

16

Lessons Learned and The Road Ahead

COVID-19 pandemic made an unthinkable impact on the businesses, industries, and governments around the globe. Starting from 2019, during 2020, and through 2021, organizations changed how they operate in many ways. Based on the spread and status of COVID-19 to date, a return to the pre-pandemic times personally and professionally seems highly unlikely. During this period, the inventory of manufacturers, distributors, retailers, and supermarkets was depleted for many items, including essential items. Customers preferred to shop online from the safety of their homes. Many organizations adjusted to the new norm with massive changes in organizational structure, number of people staffed in different roles, remote work vs working in an office, significantly lower foot traffic in physical stores, and delivery of goods by establishing new partnerships with organizations that specialize in home delivery. Wearing masks became mandatory in public places including shops, malls, supermarkets, clinics, hospitals, and offices. Most schools, colleges, movie theaters, clubs, hotels, motels, and entertainment parks etc. were partially or completely closed for extended periods of time. As the vaccines for COVID-19 were developed, tested, distributed, and started impacting the number of people diagnosed with, or dying due to COVID-19, lockdowns were lifted, and the recovery process started after almost two years of unprecedented devastation.

16.1 Search for Solutions to SCM Disruptions

Search for immediate and long-term solutions from the COVID-19 and other disasters became top priority of the governments, industries, and organizations irrespective of their size. Number of surveys were conducted to take the views of the stakeholders, actors, and experts to understand the impact of the disruptions caused by the COVID-19 and find suitable solutions. Some of the important survey studies conducted during the COVID-19 are:

i. McKinsey survey during May 15-22, 2020.

ii. ICT Study Group survey from early to mid-August 2020.

iii. Interos Survey.

iv. Resilience360 and Business Continuity Institute Survey during June and first week of July 2020.

v. CFO Survey March 9, 2020, and June 8-11, 2020.

vi. The Gartner Study during February – March 2020.

16.1.1 Survey by McKinsey (May 2020)

McKinsey, a leading consulting organization, surveyed 60 senior supply-chain executives from a variety of industries and geographies during May 15-22, 2020. The objective of the survey was to review the impact of disruptions made by the COVID-19 pandemic on the functioning of their organizations and the plans of the companies in managing the supply chains hereafter (McKinsey, 2020) (1). The purpose of the survey was to make the organizations flexible and resilient to efficiently face similar disruptions in the future.

16.1.2 Survey by ICT (August 2020)

The ICT Study Group sent a Questionnaire to companies in the ICT sector to assess the impacts of COVID-19 on ICT supply chains in terms of operational resilience and business continuity issues. The survey, conducted from early to mid-August 2020, included 50 companies ranging in size from large, publicly traded, global companies to small, and privately owned firms. The survey included 5 Broadcasters, 34 as Communications Services Providers (CSP), and 11 as IT Service Providers (ITSP).

The Study Group asked the respondents the extent the organization had been impacted by supply chain issues during the pandemic. 53% of respondents reported moderate impacts to their organizations' supply chains because of the COVID-19 pandemic, and 33% of respondents experienced minor impacts (CISA, 2020). None of the companies experienced catastrophic impacts that would have stopped their business. Going by industry segments, 60% of Broadcasters experienced moderate interruptions to their supply chain while 56% of CSPs and 40% of ITSPs also experienced moderate

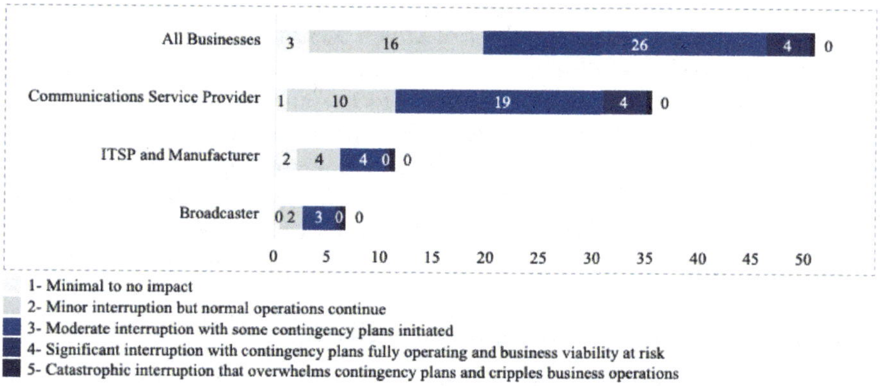

Source: CISA (2020)

Figure 16.1: Responses by Impact Rating & Industry Classification

impacts. Two of the ITSP companies informed minimal to no impact to their supply chain due to the pandemic, while only one of the CSPs and none of the Broadcasters had minimal to no impact.

16.1.3 Survey by Interos

Interos, in view of the large-scale impact of COVID-19, undertook a study titled as "Building Resilience Against Disruption: The Impact of COVID-19 on Supply Chains and How Businesses are Preparing for the Next Shock" (CISA, 2020). Interos surveyed 450 executives from the aerospace and defense, financial services, manufacturing and production, technology, energy and utilities, and other commercial sectors. The survey included Risk and compliance officers, logistics, IT, procurement, and operations executives at companies with revenue exceeding $1 billion. The survey focused on the disruptions and vulnerabilities exposed by the COVID-19, and the measures they would like to adopt to strengthen their supply chains and build greater resilience.

16.1.4 Survey by Resilience360 & BCI

Reslience360 and the Business Continuity Institute (BCI) surveyed over 350 global manufacturers and retailers from 19 sectors in 77 countries during June and first week of July 2020 (CISA, 2020). 9.6% of the respondents were from the IT sector and from the manufacturing sector 8.2%.

16.1.5 Survey by PwC

PricewaterhouseCoopers (PwC) conducted several "COVID-19 Pulse Surveys" (CISA, 2020). In its first "COVID-19 Pulse Survey", PwC approached 50 Chief Financial Officers (CFO), 80% of which were from

Fortune 1,000 companies and 44 of which were based in the U.S., with the rest in Mexico. PwC (2020) conducted the survey in the week of March 9, 2020. More than one third (34%) of CFOs stated that supply chain issues were among their top three issues in the current situation. A full 30% of the companies in PwC's survey were considering making changes to existing supply chains (PwC, March 16, 2020). PwC conducted a fourth Pulse Survey between April 20-22 observing that 56% of CFOs were planning to develop additional, alternate sourcing options for their supply chains.

16.1.6 Survey by the Gartner Group

Gartner, Inc., a leading research and consultancy organization, conducted a study "Weathering the Supply Chain Storm Survey" during February – March 2020 (CISA, 2020). The study collected data from 260 supply chain professionals from a broad range of industries, including high-tech, industrial and food & beverage. Mostly, the respondents were from North and South America and the EMEA and APAC regions. Gartner (2020) found that one third of the respondents had moved sourcing and manufacturing activities out of China or have plans to move out in the next two to three years.

16.2 Issues Observed During COVID-19

The six surveys carried out by the various agencies during the COVID-19 as discussed in Chapter 16.1 revealed number of useful information such as:

i. Weaknesses in the supply chain have been exposed during the COVID-19 period. Everybody is keen to face and tackle the problems, if it happens again (McKinsey, 2020; CISA, 2020, Interos, 2020). Interos (2020) observed that different percentage of respondents encountered different levels of disruptions mainly in logistics due to COVID-19.

ii. Encountered the problem in the supplier base (McKinsey, 2020). CISA (2020), in the case of ICT companies, noted a need to diversify supply chains to a broader array of locations away from single source/single region suppliers. Presently there is concentration of technology hardware in five places: China, Taiwan, South Korea, Japan, and USA (Figure 16.3). Intereso (2020) also found vulnerability due to concentration to single source.

iii. Gartner (2020) observed that 25% of surveyed respondents have already regionalized or localized manufacturing closer to demand. It is expected that regional supply chains can help in reducing the delays and shortages in times of disruption.

iv. Some manufacturing companies were operating on lean Inventory models, which are efficient and cost effective in normal times. However, disruptions from disasters and COVID-19 lead to risks of delays in deliveries to the customers leading to economic consequences. Operating on lean Inventory model needs change to safeguard the company and consider just-in-case inventory model (CISA, 2020).

v. Some of the tier 1 suppliers are not interested in sharing the sources of supply in tier 2 or tier 3 to avoid regulatory requirements of country-of-origin information. This issue needs attention (CISA, 2020).

vi. Many ICT companies have hundreds of tier 2, tier 3, and tier 4 suppliers but they do not know who they are? Transparency is needed to make the supply chain more resilient (CISA, 2020).

vii. Some of the governments have not identified the manufacturers of critical items.

viii. Problems were faced in the areas of production and distribution, specifically in food and consumer products (McKinsey, 2020).

ix. ICT companies need to refine their supply chain risk management strategy and control their financial burden especially from the disruptions from pandemics and disasters (CISA, 2020).

x. ICT companies do not have standardized mapping tools for effectively mapping the locations of the sub-tier suppliers and to identify upstream bottlenecks (CISA, 2020).

xi. Companies do not have alternative plans in logistics and transportation. There is a need for alternative plans to meet the demands of disruptions from different types of events. Technology use in studying the real-time needs lacks at present (CISA, 2020)

xii. SCM struggling with inefficient digital technologies (McKinsey, 2020).

xiii. Sudden remote working that made decision-making slow in planning (McKinsey, 2020).

xiv. Intereso (2020) found COVID-19 as the biggest ongoing risk followed by cyber threats.

xv. There is a need to raise the level of supply chain's resilience (McKinsey, 2020).

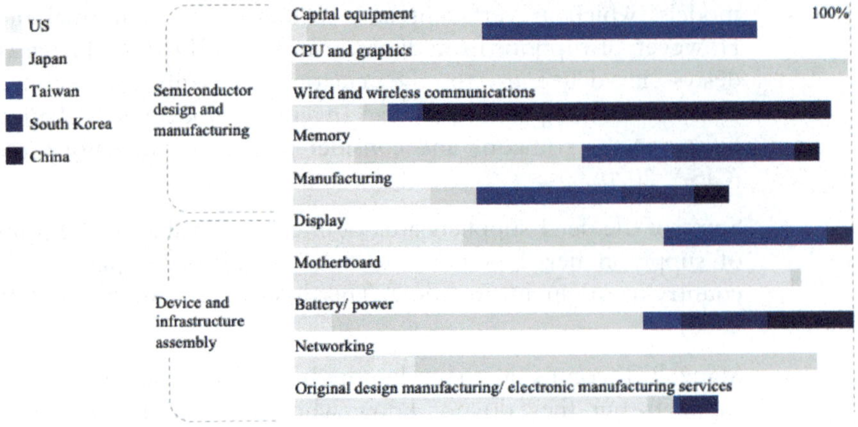

Source: CISA (2020)

Figure 16.2: Concentration of Technology Hardware in Five Countries

xvi. During the COVID-19, the Mexican government did not offer the protection of the critical item manufacturers. The governments should consider this to make the supply chain more resilient (CISA, 2020). During COVID-19 and other disruptions, companies often lacked transparency about 2nd and 3rd tier suppliers and beyond (CISA, 2020; BCI, 2019). This is a cascading effect of the failure of the 2nd and 3rd tier suppliers in cases of slowdowns, lockdowns, and disruptions. It is difficult for the company to know where exactly the delay is taking place. There is a need to map lower tier suppliers (CISA, 2020).

16.3 Human Resource Challenges

Human resource is the biggest challenge for making the supply chains resilient to manage pandemic and disasters. COVID-19 has shown that talent and technology are both key to growth (Meduri, 2014). Looking at the complexities of supply chain operations, it is essential to have a skilled workforce for effective management of the pandemic and disaster. The workforce should also be trained in humanitarian logistics encompassing three important functions: preparation, response, and reconstruction (Kovacs and Spens, 2007).

Humanitarian logistics have a great role in the response phase of the disaster. This is the priority when pandemic and disasters put their foot forward for the first time and when relief organizations, whether governmental or non-governmental including organizations-for-profit and organizations-for-nonprofit, come into action. These organizations should have competent and trained workforce. The human resources in such organizations should have a high level of strategic and process driven skill sets. In addition, they should have a high level of humanitarian logistics, planning, management, and technological skills. These skills cannot be developed in a short time but need systematic development over a period.

16.3.1 Skillset for Logistics Management

The T-shaped model of logistics skills suggested by Tatham et al. (2010) identifies four areas of skill sets that include general management, problem-solving, interpersonal, and functional logistics skills. This model lacks the most important skill set in digital technology. Digital technologies play a very important role in effective management during the pandemic and disasters. The recommended model for skill sets for personnel working in disaster and pandemic management is shown in Figure 16.4. With the large number of disasters occurring across the world, the severity of the COVID-19, new variants like delta, and Omicron shaking the whole world, it is very essential to deploy human resources of high quality. Human resources should be suitably trained with a high level of humanitarian values. It is important for the organization to train and develop competent resources. New, competent, and trained resources should be recruited to mitigate the risk to the organization due to future disasters and pandemics. Last but not the least, organizations need to evaluate the risk of losing trained and experienced resources to competition, and create an employee retention plan that:

i. Helps employees find purpose in their work

ii. Provides opportunities for career growth

iii. Provides flexibility for work-life balance

iv. Offers opportunities to take sabbatical

v. Helps adapt to employees' needs as they change

16.3.2 Impact of COVID-19 on Employees

Impact of COVID-19 on employees and employers was very strenuous. Employers were struggling to manage operations, businesses, and serve the customers' needs in the difficult situation of lockdowns and slowdowns created by the COVID-19. Industry did not need the normal strength of the workforce and made drastic cuts in the workforce on one hand, and on the other hand, it was not able to get the required number of trained manpower during the COVID-19. Travel restrictions, shortage of transport vehicles, trains, buses, cancellation of flights, and shortage of drivers made the situation very difficult. Industry found it extremely difficult to retain the trained workforce due to reduced market demand. The remote working

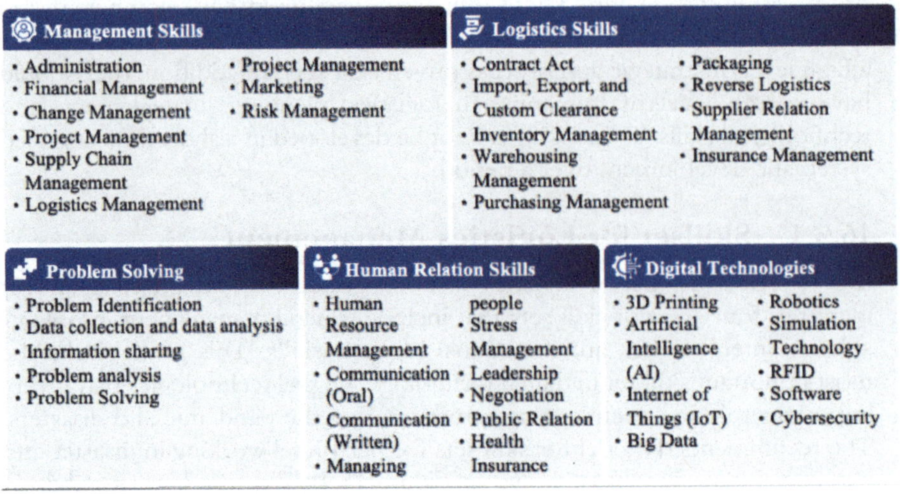

Figure 16.3: Model for Workforce Planning in Disaster

pattern has continued for about two years now. Many organizations made permanent changes in their working styles adopting remote work as a regular style of work.

Remote working created its own problems such as lack of working environment where personal interaction is minimal. Monotony of working remotely from home, and missing friends and colleagues, created psychological barriers in the working population.

Tension from the COVID restrictions and fear of job security hanging over the head led to mental tensions on the employees. Many employees proceeded to their villages and hometowns as they were unable to pay rent for their condos, apartments, or houses.

Contrary to the above scenario, many industries and businesses were facing manpower crunch due great demand and lack of availability of trained manpower. Such organizations were forced to pay higher salaries due to the poor quality of resources available. The labor market witnessed a turbulent era where employees were moving from one organization to another for a pay raise. Many employees were leaving jobs due to stress resulting from the uncertainties associated with COVID-19 in many aspects of day to day lives.

16.3.3 Voluntary Attrition

McKinsey (2021) surveyed employees and employers to understand the reasons for voluntary attrition in the labor market during the COVID-19 period. The survey covered several industries in five countries: Australia,

Canada, Singapore, the United Kingdom, and the United States. The employee survey encompassed 5,774 people of working age and the employer survey consisted of 250 managers specializing in talent such as chief talent officers, who were working with large organizations with more than $1 billion in revenues and midsize organizations with revenues from $50 million to $1 billion.

Survey found that more than 19 million US workers have left their jobs during the period April – September 2021, adversely affecting the businesses. Some companies resorted to pay hikes but did not get the solution to the problem. Some of the observations made from the employee surveys are:

i. Employees are looking for human aspects of work.

ii. They are looking for a sense of purpose in their work with social and interpersonal interaction with co-workers and superiors.

iii. They are searching for a shared identity.

iv. More than salary, perks, and benefits they are looking for being considered by their managers and the organizations as valued employees.

v. The search for autonomy and flexibility at work are haunting the employees to think of even leaving the traditional full-time job pattern. It is further noticed that employees are willing to leave a job without having another job in hand.

vi. In the turbulent labor market, even satisfied employees are tempted to leave the job due to more available opportunities with remote work options.

vii. Employees want a career path and development opportunities.

viii. In the changed environment of COVID-19, health and safety aspects are important.

Many companies are not realizing the aspects that are leading employees to leave the jobs. The element 'understanding the employees and their changing needs' is missing. The need is to introduce flexibility, connectivity, sense of unity, and purpose. The findings from all the countries and industries are consistent and more important to note is that they are likely to continue or even worsen in near future. Understanding the ground realities, some of the organizations have taken innovative measures like "staycation" packages: a movie night with popcorn and a gift-card or coffee-gift-cards to encourage connectivity.

16.4 Resilience in SCM: Looking Back

Various studies carried out by different agencies provide lessons to make the supply chains resilient to meet any demanding situation like COVID-19 and new variants like Delta-variant and Omicron-variant. Companies can take following lessons from the above studies:

i. Diversify the locations away from single source/single region suppliers to improve the resilience during pandemics.

ii. Adopt dual sourcing for raw materials.

iii. Adopt regionalized or localized manufacturing closer to demand.

iv. Increase the inventory of critical items.

v. Boost local manufacturing for critical industries that will help in more job opportunities to the local workforce (KPMG, 2020).

vi. Online shopping is a new norm in supply chains. Digital marketing, operations and sales teams use a single platform along with 24/7 training, sales forecasting, and analysis of customer data. The advantage is real-time visibility that supports improved decision making and reduced operational cost.

vii. Adopt nearshoring and increasing supplier base.

viii. Change over from operating on lean-inventory-model or Just-in-Time (JIT) model to Just-in-Case (JIC) inventory model. This will help them to adhere to committed delivery of their product to the customers in times of disasters and pandemics. Customers will get economically protected with continued delivery of supplies. Although the JIC inventory model will slightly increase the inventory levels and thereby inventory-carrying-costs, the model will enable companies to deliver the materials on time during emergency periods and create better resilience for the supply chains.

ix. Map the 2nd tier, 3rd tier, and lower tier suppliers so that cause of delay in case of slowdowns, lockdowns, and disruptions may be identified and controlled. This will provide better resilience to the supply chains.

x. Develop alternative plans in logistics and transportation to meet the demands of disruptions from different types of events. Technology should be used in studying real-time needs. Alternative plans in logistics and transportation in the areas of production and distribution, food, and consumer products need immediate attention.

xi. Refine supply chain risk management strategy and control the financial burden especially from the disruptions from pandemics and disasters.

xii. Develop mapping tools for effectively mapping the locations of the sub-tier suppliers and to identify upstream bottlenecks.

xiii. Supply Chains should do away the inefficient digital technologies. Efficient digital technologies offer improved resilience to the supply chains.

xiv. Adopt customer centric, data enabled, digital technology. Investment in automation, digital enablers and advanced technologies offer lower cost of production due to reduced labor inputs (KPMG, 2020).

xv. Train the workforce to develop skills within the organization in using new digital technologies. Skilled workforce should be capable of using the new digital tools efficiently.

xvi. Develop skills set for prompt response to the pandemics and disasters. Skills set for humanitarian logistics should be focused and developed.

xvii. Hire new skilled workforce for building teams.

xviii. Make changes in the planning processes on a permanent basis, which may include increasing the centralization of planning activities, short planning cycle, and adopting advanced-analytics techniques.

xix. Introduce long term changes keeping in view the needs of customers and their behaviors like online purchasing.

xx. Working remotely from the home, learning while staying-at-home, order from the comfort of home and interacting with the whole world staying at home has given a push to digital technologies in supply chain management. This change in work style is going to continue in future as it offers a number of side benefits like saving in travel time, less burden on infrastructure, saving in fuels and many other benefits.

xxi. The benefits of technology adoption have been fully understood during pandemic COVID-19. Real-time decision-making and responsiveness have changed the business world. Adopt technologies to monitor and adjust operations according to change in customer behavior and flexible character of supply chains.

xxii. Remove humans from routine tasks by using robotics, IoT, and sensors. Adopt Artificial Intelligence (AI) and analytics to collect and analyze internal/external data for better and fast decision making. Adoption of these technologies will create a different working environment with different skillsets and economy of operations.

xxiii. Adopt advanced tracking technology to track and trace the movements of the raw materials and finished goods. Pharmaceutical supply chain has benefited tremendously in COVID-19 period from these technologies. Walgreen and other pharmaceutical companies in the USA are effectively using tracking and tracing the supplies and making their supply chain economical and effective.

xxiv. Adopt data analytics, machine learning (ML), and other technologies to predict supply chain disruptions. This is very relevant in case of pandemics and major disasters. These technologies will help in operations planning, inventory management, production planning, and risk analysis. Simulation technology can further help in optimization efforts. Strategic and concentrated supplies can be managed more efficiently by adopting these tools.

xxv. Review and modify the organization structures by integrating the remote and normal on-site work. Tasks that can be done remotely without sacrificing the efficiency, should be identified in advance. A balanced work plan should be developed for efficient operations. Planning process should be made agile to make decisions faster in times of emergency.

xxvi. Invest in Cognitive Decision Centers (CDCs), which capture and interpret the cross-functional data to remove conflicting priorities and help in taking strategic decisions to trade-off and make better decisions supported by AI. This helps in improving the performance of the whole supply chain.

xxvii. COVID-19 is expected to continue in different forms due to new variants appearing from time to time from different countries in the world, specifically South African and Asian countries. Similarly, cyber threats are faced from different parts of the globe, difficult to identify. Companies need to develop strategies to protect themselves from pandemics and cyber-attacks.

xxviii. The supply chain resilience is linked with the preparedness of the company. Every company should assess the criticality of its operations and improve safety of the supply chains appropriately with adequate resources and funds. Experience of COVID-19 and recent cyber-attacks reported from different parts of the world call for strengthening the supply chain resilience.

xxix. Adopt smart investment rather than just investing more money in the supply chain.

xxx. Consider retaining the environment benefits like low level of pollution from low level of activities.

16.5 Resilience in SCM: Looking Forward

Road ahead for the next '**normal**' is not easy but can be made approachable by proper analysis, planning, implementation, and control. All the stakeholders must jointly work together to achieve a common goal to fight the COVID-19 and similar threats from disasters. A proper analysis of the issues is necessary for bringing the next normal, followed by strategic action plans for short-term, medium-term, and long-term.

16.5.1 Key Considerations in Strategy Development

COVID-19 has created an environment of fear in everybody's mind for almost the last two years. After the development of vaccines and two shots for protection, people were relaxed to some extent, but the new variant 'Omicron' has come up and has already reached 59 countries in the world, the number is increasing every day. Omicron is more deadly than COVID-19 first and second waves and spreads much faster. Companies are struggling with the plans to reach the next 'Normal'. The path is having many concerns and it is difficult for anyone to write a plan for the return of normalcy even though vaccines for the treatment are now available in adequate quantity, and herd immunity has been reached.

i. **Remote Workforce**: The second concern about the new working style 'remote working'. Although many companies were happy that a new working style has successfully emerged and would solve problems like COVID-19, its adoption is still very low. McKinsey (2020) (2), in its research, found that at 15 top firms in Silicon Valley, only 8 percent of the employees work remotely. These firms are not very keen to continue with this new norm of work. A balanced working style is needed to manage operations without any hurdles during disruptions due to disasters and pandemics.

ii. **Exponential Growth in e-Commerce**: The third uncertainty is in customer's new preference for online shopping, forced by the lockdowns and slowdowns. Online business grew very fast with growth of up to 60 percent in some categories (McKinsey, 2020), but what shape it will take in future is anybody's guess. There are many more areas, which need to be addressed to deal with the challenges ahead. McKinsey (2020) (2) has outlined four forces: demand planning, changes in the workforce, new regulations from the governments, and information about protocols for safety.

iii. **Demand Planning:** Type and level of demand is yet to be stabilized. The change in online shoppers is quite visible over the time. About 20-25 percent of people are now using digital channels. With the increase in percentage of digital customers, the customer loyalties for brands are also changing. The market structure is taking new shape. For example, the number of customers for mobile banking is increasing fast. Similar is the case in the education sector where online classes are on the increase. Demand planning has become more complex than ever. Different sectors have different scenarios of demand recovery.

iv. **Rapid Changes in Workforce:** COVID-19 has affected the economy very badly in almost all the sectors all over the world. The impact has been different in different sectors and countries based on the severity of the COVID-19. McKinsey (2020) (2) estimates that one-third jobs in the USA may be vulnerable from pay-cuts and lay-offs. Eighty percent of those jobs fall in the low category workforce. Many of the employees are unable to join their duties due to health issues as some of them are sick or quarantined. One of the most important reasons for this is the shortage of employees with the needed skills in various areas of activities.

v. **Changes in Social Behavior**: Remote working has its own problems. People miss the office environment which they have been accustomed to for decades. Others do not find home-office setup in remote working. Of course, there is some percentage of employees who are happy to work remotely as they save time commuting long distances every day and can produce more work at the same time. Further they enjoy the comfort of working at home while working remotely. The conflicting likes and dislikes of employees is to be resolved by the employers by providing a balanced work solution.

vi. **New Regulations**: Various countries have taken a variety of measures to control COVID spread. The new variant "Omicron' has created a lot of fear in different countries. New stricter measures are introduced including the social distancing norms and wearing of masks. Some countries have made vaccination's three doses compulsory and entry to restaurants, entertainment places and other places of gathering is banned without negative COVID tests. In the United States, some states have imposed a ban whereas in some other states, the ban has been lifted or relaxed. In democratic set ups like the USA and India, the state governments in some of the states are in the control of opposition parties, both national level parties and the regional parties. Due to the prevailing political system, different states have different regulations. Supply chain resilience in such a system has a different degree of application.

vii. **Information System**: There is a need for a fast, accurate, and trustworthy information system to provide correct information in a simple way. TV programs and mobiles are doing a good job in information dissemination. The supply chain resilience is greatly helped with information sharing. Information about testing facilities, availability of vaccines can help companies to develop proper guidelines about social distancing, stagger shifts, and hybrid onsite/remote models. These efforts are going to create resilient supply chains.

16.5.2 Strategic Action Plans for Resilience in SCM

A good starting point for developing strategic action plans is to use the lessons learned during the COVID-19 period as discussed in Chapter 16.4. All the lessons should be critically analyzed and classified to develop short-term, medium-term, and long-term plans.

i. **Short-Term Strategic Action Plan**: Short-term strategic action plan is expected to be executed within one year to enhance the resilience of supply chains. Various actions under short-term plan include the following:

   - Adopt multiple sourcing for raw materials to meet the disruptions resulting from pandemics. Source of supply should be diversified away from single source/single region suppliers to improve the resilience during pandemics. In addition, the supplier base should be increased.

   - Map the 2nd tier, 3rd tier, and lower tier suppliers so that cause of delay in case of slowdowns, lockdowns, and disruptions may be identified and controlled. Develop mapping tools for effectively mapping the locations of the sub-tier suppliers and to identify upstream bottlenecks. This will provide better resilience to the supply chains

   - Adopt regionalized or localized manufacturing closer to demand. Boost local manufacturing for critical industries that will help in more job opportunities to the local workforce (KPMG, 2020).

   - Increase the inventory of critical items by considering Just-in-Case (JIC) model in place Just-in-Time (JIT) model used in normal conditions for economizing the cost.

- Online shopping should be encouraged to the extent possible to help in minimizing personal visits by the customers and reduce exposure to COVID attacks. Online shopping is also economical, convenient, and time saving.

- Develop alternative plans in logistics and transportation to meet the demand from different types of events at the time of disruption. Technology should be used in studying real-time needs. Alternative plans in logistics and transportation in the areas of production and distribution, food, and consumer products need immediate attention.

- Refine supply chain risk management strategy and control the financial burden especially from the disruptions from pandemics and disasters.

- Organize training programs to develop skills within the organization in humanitarian logistics, analytics and in using new digital technologies. Skilled workforce should be capable of using the new digital tools efficiently. Skills set for prompt response to the pandemics and disasters should be developed.

- Hire new skilled workforce for building teams and to fill up the gaps in knowledge.

- Planning processes should be centralized and changed on a permanent basis to bring in improvements, short planning cycles, and adopting advanced-analytics techniques.

- Working remotely from the home, learning while staying-at-home, ordering from the comfort of home and interacting with the whole world staying at home has given better resilience to supply chains. COVID-19 forced the organizations to work from home. This change in work style is going to continue in future as it offers a number of side benefits like saving in travel time, less burden on infrastructure, saving in fuels and many other benefits.

- Companies need to develop suitable strategies to protect themselves from pandemics and cyber-attacks.

- The supply chain resilience is linked with the preparedness of the company to respond to the pandemic. Every company should assess the criticality of its operations and improve safety of the supply chains appropriately with adequate resources and funds. Safety measures should be strengthened to meet the challenge of pandemics and cyber-attacks.

ii. **Medium-Term Strategic Action Plan**

- Supply Chains should do away the inefficient digital technologies. Efficient digital technologies offer improved resilience to the supply chains.
- Induction of technology removes humans from routine tasks. For example, using robotics, IoT, and sensors helps in doing many operations without dependence on human beings. Adoption of technologies will help improve working efficiency that is very much needed at the occurrence of pandemics and disasters.
- Selection of production sites with the economy of labor cost and availability of raw materials will help building resilience in supply chains.
- Online shopping should be expanded. Digital marketing, operations and sales teams use a single platform along with 24/7 training, sales forecasting, and analysis of customer data. The advantage is real-time visibility that supports improved decision making and reduced operational cost.
- Make continuous efforts in developing the skills set for prompt response to the pandemics and disasters. Skills set improvement takes a long time and requires support from top management.

iii. **Long-Term Strategic Action Plan**

- Digital technologies have brought many advantages to supply chain management. Artificial Intelligence (AI), Simulation, big data, robotics, 3D printing, and Internet of Things (IoT) have contributed to a great extent in supply chain management and increasing the supply chain resilience. The benefits of technology adoption have been fully understood during pandemic COVID-19. Real-time decision-making and responsiveness have changed the business world. Adopt technologies suitable to individual cases. Use of a combination of technologies has given much more benefits. Selection of technologies should be done carefully after detailed analysis.

- Adopt advanced tracking and tracing technology to monitor movements of the raw materials and finished goods. Pharmaceutical supply chain has benefited tremendously in COVID-19 period from these technologies. Walgreen and other pharmaceutical companies in the USA are effectively tracking and tracing the supply of medicines and making the supply chain economical, effective, and resilient.

- Adopt data analytics, machine learning (ML), and other technologies to predict supply chain disruptions. This is very relevant in case of pandemics and major disasters. These technologies will help in operations planning, inventory management, production planning, and risk analysis.

- Review and modify the organization structures by integrating the remote and normal on-site work. Tasks that can be done remotely without sacrificing the efficiency, should be identified in advance. A balanced work plan should be developed for efficient operations. Planning process should be made agile to make decisions faster in times of emergency.

- Invest in Cognitive Decision Centers (CDCs), which capture and interpret the cross-functional data to remove conflicting priorities and help in taking strategic decisions to trade-off and make better decisions supported by AI. This helps in improving the performance of the whole supply chain.

- Adopt customer centric, data enabled, digital technology. Investment in automation, digital enablers and advanced technologies offer lower cost of production due to reduced labor inputs (KPMG, 2020).

- Introduce long term changes keeping in view the needs of customers and their behaviors like online purchasing.

- The benefits of technology adoption have been fully understood during pandemic COVID-19. Real-time decision-making and responsiveness have changed the business world. Adopt technologies to monitor and adjust operations according to change in customer behavior and flexible character of supply chains.

- Remove humans from routine tasks by using robotics, IoT, and sensors. Adopt Artificial Intelligence (AI) and analytics to collect and analyze internal/external data for better and fast decision making. Adoption of these technologies will create a different working environment with different skills and economy of operations.

- Adopt advanced tracking technology to track and trace the movements of the raw materials and finished goods. Pharmaceutical supply chain has benefited tremendously in COVID-19 period from these technologies. Walgreen and other pharmaceutical companies in the USA are effectively using tracking and tracing the supplies and making their supply chain economical and effective.
- Adopt data analytics, machine learning (ML), and other technologies to predict supply chain disruptions. This is very relevant in case of pandemics and major disasters. These technologies will help in operations planning, inventory management, production planning, and risk analysis. Simulation technology can further help in optimization efforts. Strategic and concentrated supplies can be managed more efficiently by adopting these tools.
- Review and modify the organization structures by integrating the remote and normal on-site work. Tasks that can be done remotely without sacrificing the efficiency, should be identified in advance. A balanced work plan should be developed for efficient operations. Planning process should be made agile to make decisions faster in times of emergency.
- Invest in Cognitive Decision Centers (CDCs), which capture and interpret the cross-functional data to remove conflicting priorities and help in taking strategic decisions to trade-off and make better decisions supported by AI. This helps in improving the performance of the whole supply chain.
- Develop appropriate strategy to protect from pandemics and cyber-attacks.
- Adopt smart investment rather than just investing more money in the supply chain.
- Consider retaining the environment benefits like low level of pollution from low level of activities.

16.6 Summary

COVID-19 has shaken the whole world with large scale deaths, lockdowns, slowdown, and various restrictions imposed by the governments of various affected countries. This Chapter is one of the most important Chapters of this book. Large scale loss of lives, damage to economy, inconvenience caused to the public, and loss of revenue to the business led to number of research studies by some of the reputed research organizations that included McKinsey, ITC Study Group, Interos, Resilience360 and Business Continuity Institute, CFO, and Gartner. A consolidated list of issues observed in the six research studies is presented. In addition to the issues observed in the six research studies, the Chapter deals with human resource challenges in supply chain resilience. COVID-19 impacted the employees and employers of various organizations during the period of COVID-19. Special focus is on the employees and what skill sets are essential for the employees dealing with the pandemics and disasters. The reasons of employee's attrition have been discussed in detail. Chapter then deals with the lessons learned from the findings of the research studies.

Road ahead for the next 'normal' after COVID-19 is the primary focus of the Chapter. After deliberating on the main issues in developing strategy to bring the next normal, the chapter deals with the barriers in developing the strategy. Finally, the strategy worked out has three action plans: the short-term action plan, medium-term, and long-term plans. These plans have listed several actions that should be taken to manage the pandemics and disasters. It is hoped that the strategic action plans will provide guidance on how to improve the resilience of the supply chain to deal with the difficult situation of pandemic or disaster.

References

CISA (2020). Building a more resilient ICT Supply Chain: Lessons learned during the COVID-19 pandemic. Cybersecurity and Infrastructure Security Agency. 1-52.

Gartner (2021). Gartner's Disruption Management Survey Reveals What Differentiates Fit from Fragile Supply Chains. Gartner 16 March 2021.

Kovacs, G & Spens, K M (2007). Identifying Challenges In Humanitarian Logistics. International Journal Of Physical Distribution And Logistics Management. 39 (6).

KPMG (2020). Building Supply Chain resilience through digital transformation. KPMG International. 1-6.

McKinsey (2020) (1). Resetting supply chains for the next normal. McKinsey and Company. 1-7+.

McKinsey (2020) (2). Return: A new muscle, not just a plan. McKinsey and Company. 1-6.

McKinsey (2021). Resilience: The whole-company fitness challenge. McKinsey & Company. 1-6.

Meduri, Yamini (2014). Humanitarian Logistics: Challenges for Human Resource Management. Universal Journal of Industrial and Business Management. 2 (6). 135-141.

Tatham, P., Kovacs, G. &Larson, P. (2010). What Skills and Attributes Are Needed by Humanitarian Logisticians-A Perspective Drawn from International Disaster Relief Agencies? PPMS 21stAnnual Conference, Vancouver, Canada.

INDEX

Made in United States
Orlando, FL
07 November 2026

90337545R00115